TOWARDS ETERNITY

Bijay Banthia

RoseDog Books
PITTSBURGH, PENNSYLVANIA 15238

RoseDog Books
585 Alpha Drive
Suite 103
Pittsburgh, PA 15238
Visit our website at *www.rosedogbookstore.com*

ISBN: 979-8-88925-058-6
eISBN: 979-8-88925-558-1

PREFACE

Life sometimes takes such an unusual turn that it is neither possible to forecast it, nor explain it. This is almost a true story with a little fiction of a super-rich person relinquishing everything including his family. He was dragged into some ghostly and unimaginable circumstances and fought hard for years and years. Fraud and manipulations by his own did not have any end. Suddenly, he saw light in giving up everything and taking a noble route, which inspired all around him. He attained supreme peace, a level that he could not achieve even if he possessed the entire wealth of the world.

CONTENTS

TOWARDS ETERNITY

I.

SCHOOL FRIENDSHIP

September 15, 1972

Bijay just returned home to Kolkata, India, after about 10 years' stay in Dallas, Texas, USA. It was middle of the night at the Dum Dum Airport, and on picking up his baggage, he took a pre-paid taxi to go to his parents' residence in Bhowanipur near Calcutta Club. When he was riding the cab, it passed through Ras Behari Avenue in Ballygunj suburb, where his long-time friend Sajjan lived. Memory took him several years back, and though he was extremely tired due to a very long journey from USA to India, and almost half-sleeping, he could not forget his time with Sajjan in school and thereafter. Though he knew that he would be in Kolkata for a short while only, he decided to call him, knowing fully well that he might not even remember or recognize him, and care for him because of time gap and no communication with him whatsoever. Besides, Sajjan was too rich and well-off to care for an ordinary friend like Bijay.

Bijay felt sorry that he had to leave Kolkata about ten years back in a great hurry. He could not even inform his dear and near ones, including some of his long-time friends like Sajjan. This was because he got last minute acceptance for admission into the University of California at Berkeley, California, USA. He was asked to complete a long list of unimaginable formalities before leaving, including applying for a student visa for USA, for which vaccinations and several medical examinations were required. His very last contact with Sajjan and others was when he was still trying to get a job in India on completing his undergraduate Engineering Degree. Though Sajjan was more than capable of hiring him in his father's big business empire, Bijay knew Sajjan could not use his Engineering skills, and it would not be a well-paid job in any case. It would just be a favor to Bijay, and for sure he did not wish to be a burden on his friend. This is also true that Bijay never approached him, and he was still hopeful that his completing the engineering degree with distinction and first rank in class would eventually help him in getting a job without any body's help. It never materialized, as he did not have high-level contacts with companies in India. Some of his friends in similar situation went overseas, so he also tried his luck for admission into a US University and getting some financial assistance.

University of California's application process was so complicated with so many requirements that he almost gave up. He had, however, no other choice as he did not get any job in his profession despite approaching all relevant companies in India. He hated to while away time sitting at home doing nothing. Though his parents did not say anything to him, he felt guilty within himself. Good part was that he got good references from his engineering school. He was always hopeful that eventually his ranking etc. would count. The application process required all educational transcripts to be notarized. Somehow, he filled the form and sent to the University, which promptly answered back. Though he was accepted for admission, and a part-time Research Assistantship, he had a deadline for registration in just 2 weeks with tons of formalities to be completed regarding obtaining passport (he never had one so far), US Visa, travel arrangement etc. By the time, he completed these, he did not have any time to inform about his plans to leave for USA even to his closest friends like Sajjan etc.

Bijay reached home and decided to have a nice sleep, as he was very tired after almost a twenty hours' long air journey from USA to India. It was time change or jet leg also, and probably it would take several days before he would be normal again. He realized that it was best for him for the time being to forget about Sajjan and others and enjoy his stay with parents and relatives. He felt sorry that he did not keep communication even with his close friends like Sajjan, as he was over-burdened with enormous pressure from higher studies and strict deadlines on work, projects etc. in USA. Also, he had to face 'culture shock' for several weeks when he first reached USA.

Getting Normal

It took several days before Bijay recovered from the jet leg. He unpacked his stuff, distributed gifts that he had brought with him for each family member, who really appreciated his affection for them. This was, in fact, nothing unusual. It was a trend in those days that those who did well academically, often aspired to go abroad for higher studies, usually graduate studies. Part of the reason was that it was very difficult to get a job in India unless you had contacts with very high ups in big companies. Secondly, USA was one country, which was still granting educational visas for students in good standing. Some were fortunate to be awarded financial support in the form of

Research Assistantships etc. It was hard for Bijay to understand that such people were in short supply in USA and need was much greater there. He often wondered; USA must be having her own people- why should they care for importing people from abroad, specially knowing nothing about them? It took him 4 long years to complete the degree in India, spending a lot of time and money, yet no job! Though normally they were expected to return to India on completing their studies, but most of them applied for jobs in USA and stayed back with work visas. This was the case with Bijay also. He not only got his student (I-20) visa, but later got immigration visa (green card) and subsequently US Citizenship.

Recollecting School days

Bijay was relaxing at home. Suddenly, he was reminded about his long-time friend, Sajjan. He remembered to have passed through his house in the middle of night in the cab and wondered about his well-being. He decided to call him and see him at his convenience, if he had time to spare for Bijay. He was very aware that his friend must be extremely busy in his business empire and might not even pick up his phone. Even his secretary could turn it down, as Bijay had no business connection with him.

Those were the days when he happened to be his classmate in a local high school. Every morning, Bijay used to walk through very crowded narrow streets in the Bara Bazar (Big market) Area of Kolkata. In the evening, when the school was over, he used to pack his bag with class notes, books, lunch box etc. As he used to get ready to walk back home, one of his twenty classmates, Sajjan, asked Bijay to ride with him in his car. Bijay had never talked to him, as classes kept both Sajjan and Bijay very busy. In the lunch hour (only spare time in the school), Bijay quickly used to take his home-made lunch in a corner of the classroom. He could not see Sajjan in the classroom during lunch time, and he did not have any idea about his place of lunch, and never had a chance to talk to him. He convincingly told Bijay that dropping him was no big deal for his driver, as Bijay's rented residence (a small room) was on his way.

Bijay, first hesitated, as he did not know Sajjan that well, and further he did not wish to inconvenience him or anybody, as walking back and forth from school to home and home to school was his normal routine. Any way, he accepted Sajjan's offer, which he made in a very friendly tone. Bijay was astonished to see his big and neat car, and his driver in formal dress with hat.

In fact, Bijay felt hesitant and embarrassed, to step into his car with his ordinary bag, dress and dirty shoes, and smartly gave his address to the driver, who dropped him apparently on way to Sajjan's home. Though Sajjan was in a talking mood, Bijay was completely lost in the comfort of the ride vis-a vis his struggle to walk through very crowded streets.

First, Bijay thought it was just a one- time offer, but when Sajjan was courteous enough to give a ride most often, Bijay not only felt grateful to him, but he also told his parents about his friend giving ride to him. His mom was so pleased with his friend Sajjan that she asked Bijay to invite him sometime for lunch at home. Bijay's father and Bijay both felt a bit hesitant to invite him in their small room, but Bijay's mom wanted to express the family's gratitude to him by preparing meals by her own hands and serving food herself. She did not care about her tiny room with no furniture, as for her, Sajjan was just like her own son, Bijay. Who could be that nice to Bijay as Sajjan was? The family did not have anything worthwhile to give to him as a gift otherwise, and so she insisted Bijay to invite him one day. She did not realize that there was no parking space in front of her house. Also, Sajjan was too fat and heavy to take the stairs to the fifth floor, where she lived in a small room with her family. There was no elevator, and the stairs were rough.

Bijay had no idea about Sajjan's house. It was so far, and Bijay never went that far, even though he had been living in Kolkata since many years. He wondered as to why Sajjan would accept such a mediocre high school so far from his house. Obviously, Sajjan must be very well-off to afford such a luxury car with a well-trained professional driver. He could probably go to a posh school, possibly near his house.

Bijay became very anxious to find out more about Sajjan. Why was he so kind to Bijay? Would his parents not mind if they come to know about his giving ride to Bijay? He had so many questions in his mind, but did not have courage to ask his friend, who was so nice to him. After all, why should he ask? Sajjan never asked for anything in return from Bijay. He did not know that this school was near his father's business office, nor did he know that Sajjan appreciated Bijay's performance in the class. Bijay, every year, was at the top of his class in each subject and Sajjan, at the back of his mind, had admiration for him. He never expressed his feelings to Bijay and kept high opinion about him in his mind.

On repeated insistence from his mom, Bijay finally mustered strength and asked his friend to stop by for dinner one day at his convenience.

Sajjan agreed immediately, and though it was only early evening, he got down from the car and asked his driver to park somewhere and come back in an hour. Bijay knew that his mom was not at all prepared to serve them dinner so fast, but he mustered the strength to take him to his room on the fifth floor. He shouted from the basement of the house and told his mom that Sajjan was coming for dinner, and he would be there only for an hour.

Bijay's mom quickly lit her coal-fired stove, prepared two vegetable curries, Raita (churned yogurt with minced cucumber), mango pulp and hot flat breads quickly. Bijay made it a point to follow Sajjan while going up the stairs, as Sajjan was having a hard time and was almost out of breath. Bijay politely asked him to take it easy and take as much time as needed. Bijay quickly went up the stairs to his room to fetch a glass of water, which was not cold, as they did not have a refrigerator. Still, Sajjan appreciated taking sips of water and finally made it slowly to the fifth floor.

Bijay knew that his father would not come home before dark in the evening, and it would not be fair to Sajjan to wait for him. Had he been informed earlier, he too would have liked to meet Sajjan, as according to him, such people indeed were very rare to find. Bijay's mom, however, filled up all the gap. On preparing simple food, she served dinner for both friends, and sat right in front with a peacock-feathered hand fan to blow some air as they ate sitting on the floor. She herself brought water for them in tumblers. Sajjan was wonder struck to see this hospitality and affection. Besides to him, the food was so delicious that he asked for several scores and had a hearty meal. He was glad that he came there and thanked Bijay's mom for everything. She asked him to visit again and as often as he could come.

Bijay did not know that Sajjan lost his mother when he was still a child. Meals were prepared at his house by chefs and were served mechanically without anybody giving him company. He rarely saw his father, as he was always busy at work and never had time to say even hello to his only son, though he loved him very much.

This friendship grew day by day, but Bijay was still hesitant to ask Sajjan about his many questions that Bijay had in his mind. He made it a point not to ask for a ride back home, but never refused, if Sajjan insisted. They had different desks in the class, as they were assigned by the teacher. Bijay thought many times that sitting on the wooden chair the whole day must be very uncomfortable for Sajjan, as his car was so comfortable. Why was he taking so much hardship, if he was so well-off? It so happened that while he dropped Bijay one evening, he himself asked Bijay if he could have dinner again. It was indeed a welcome news for Bijay.

On seeing Sajjan again, Bijay's mom was overwhelmed with pleasure and started cooking dinner immediately. Luckily, Bijay's father came early from work that day, as he had some work at home. He was very pleased to see Sajjan and thanked him for being so nice to Bijay. While food was being cooked, Sajjan and Bijay did not want to waste time. Instead, they completed their homework for the next day. As usual, Bijay had taken extensive notes in the class, and these helped both. Sajjan realized that he did not pay that much attention in the class and appreciated his friend sharing with him his class notes. Seeing that Bijay was so thorough in his work, Sajjan started borrowing his notes very often and used freely. Bijay felt good that he was helpful to his friend.

That evening also, Bijay's mom served food to all and Sajjan enjoyed his dinner thoroughly once again. She asked him about his family.

Sajjan told her that he lost his mother when he was still a child and had a younger sister and father at home. When Bijay's father asked him about his father's work, he politely mentioned that he was managing family's business. He did not mention at that time that he was a business tycoon and 'Mica King of India' with mica mines in Bihar, about 200 miles from Kolkata.

Sajjan also did not talk about his father's office being in Kolkata's most prestigious area, India Exchange. By coincidence, Bijay's father also worked as a jute broker not too far from the office of Sajjan's father. Bijay's father did not have any office or fixed room. It was just a common area, where he could talk to his fellow brokers and got business news about his profession. As a matter of fact, he had just a meagre income and was somehow supporting his family of his wife and five children, including Bijay.

This was the first time Bijay knew a little bit about Sajjan, but he never paid too much attention. His world was merely studies, and his own family.

Time rolled on. When the school was going to close for summer for about four weeks, Sajjan once asked Bijay to come over to his house some time, and he asked him to call him before coming. He gave his address and phone number to Bijay. Of course, Sajjan knew that Bijay had no transport and no telephone. As the date was still unknown and uncertain, he could not arrange right now for his driver to pick up Bijay. Also, he thought that when Bijay would inform him about his coming to his house, he would easily arrange transport for him at that time.

Kolkata in Summer

Kolkata gets very hot in summer, and temperatures usually reach about 100- 105 degrees Fahrenheit with unbearable humidity in the afternoons. In fact, it gets hot right from the morning. Bijay remembered the cozy ride in Sajjan's air-conditioned car, and he wished he could afford one in his lifetime. It was just a wishful thinking on his part, as his parents indeed had a hand-to- mouth life with so many family members, including his parents, and only one hand, his father, to earn. Though his father used to work very hard, he was just in jute brokerage business with ups and downs depending on the number of accepted quotations based on lowest price of jute.

Anyway, one Sunday, Bijay decided to see Sajjan at his house. As there was no telephone in the house, he had to go to a public telephone booth to call him. Sajjan smartly picked up the telephone and asked Bijay to come on. Bijay was too hesitant to ask Sajjan to arrange for transport. This, being Sunday, the driver was off, and the car was in the garage. Sajjan was still too young to get a driver's license. The only person in the house, who could drive, was his father, and Sajjan could never ask him to pick up his friend. It was a good distance between the Burrabazar (Big Market) and Ballygunj areas. Bijay used to live in the former and Sajjan in the latter.

So here was the dilemma. How to go to Sajjan's house? Bijay could not afford taxies, which were very expensive indeed, and very difficult to get. He did not know any other way other than walking this far (about 10 miles) in this hot weather or to go by trams. He mustered the strength to ask his father and came to know that there was no direct tram going that far. He had to take one going from his place to Esplanade, and then another tram from Esplanade to Ballygunj. The other alternative was to go by bus and his father did not know the bus number. During all this confusion, Bijay decided to take the trams. In fact, this was the first time ever since he came to Kolkata that he took such a long tram ride by himself. Somehow, he reached Ballygunj tram depot, and had to walk a lot on a really very hot day to reach Sajjan's house.

When Bijay reached Sajjan's house, as per the address given by Sajjan, he was astonished to see the gates closed with an armed security guard in military uniform. Bijay took the courage to ask him if he was at the right place. Sajjan forgot to inform the security guard that he was expecting his friend, Bijay. The guard rudely but rightly refused to open the gates and allow Bijay to go to his friend's house. Bijay had no other option except calling Sajjan again from a near-by telephone booth.

Sajjan, kind of, felt sorry for Bijay and promptly asked one of his servants to instruct the gate keeper to open the doors and fetch Bijay to his house. When Bijay was nervously getting close to Sajjan's house, he saw a car passing by him. The person in that luxury car was an elderly man, very nicely dressed, and Bijay did not realize that he was Sajjan's father. When he asked, Bijay politely introduced himself and mentioned about seeing Sajjan. The elderly gentleman, in a rather hostile tone, asked Bijay as to how he knew Sajjan, and as to how he managed to enter his securely protected area. Admittedly, Bijay was scared, and was just about ready to drop the idea of seeing his friend in his house.

Right at that time, by stroke of luck, Sajjan walked out of his palatial house and greeted Bijay. What a difference? He also told Bijay that he just passed through his father in the car. Bijay felt a little sorry that he could have come a little bit earlier, and in that case, probably he did not have to see Sajjan's father in such a rough mood. Probably, Sajjan would have introduced Bijay as his friend, but it was too late now! Later, he came to know from Sajjan that his father would not be back till late in the evening, and in any case, he did not see any good reason for Bijay to wait for him for that long!

Bijay could see that Sajjan's house was on the ground floor itself, and no wonder he never ever stepped up on the stairs. The driver used to drop him right in front of his house, and he happily walked to his private palatial room with a lot of furniture, pictures, art pieces etc., which Bijay had never seen in his life so far. The two friends were indeed very far from each other in terms of their lifestyle. It is strange that Bijay never thought of missing anything till he saw his friend's house and felt extremely embarrassed to have invited him to his rented place on his mom's insistence.

Sajjan took Bijay inside his guest room. Bijay had never seen such a magnificently decorated and very spacious room with very high ceilings, fans, air conditioners, very comfortable chairs, tables, couch, beautiful paintings etc. He was simply spell-bound and could not even dream about them. In the back of his mind, he felt indebted to his friend Sajjan to have invited him to his palatial house. Sajjan never asked Bijay about the hurdles he faced in coming to his house, nor did Bijay feel any reason to tell him about his rather hard journey by trams, walking etc. in very hot and humid weather. Though he was sweating, when he entered the house, due to sweltering heat outside, he felt very comfortable inside the house, which was centrally air-conditioned.

When Sajjan asked Bijay 'what will you like to drink?', Bijay could not even understand his question, as he just knew about water and nothing else. So, he politely said 'water'. Bijay never knew that a drink is offered before meals as a courtesy. Sajjan pressed the bell and one of his attendants appeared promptly before Sajjan and saluted him. Sajjan asked him to tell the chef to serve food for 2 persons-Bijay and himself.

Bijay was surprised by the amenities that Sajjan enjoyed and yet he behaved like an ordinary person in the school. Why did he select that far-off ordinary school? Bijay did not have enough courage to ask Sajjan, but he later came to know that he did not get admission into the near-by prestigious, but very expensive Saint Xavier's School due to some unknown reasons, and this school was near his father's office. After dropping Sajjan to school, the driver used to take the car to the office and follow his father's instructions.

Lunch was served on the dining table in beautiful plates with curries and yogurt in separate bowls. Though it was entirely a vegetarian meal, as Sajjan knew that Bijay was a strict vegetarian, it had many delicacies. Bijay, a sort of felt sorry and embarrassed that he could not offer Sajjan any of these, yet Sajjan enjoyed. This certainly convinced Bijay about his friend's simplicity and good nature. During the sumptuous meals, sweets and other delicacies like ice cream etc. were also served. Bijay had a hearty meal, and decided to tell his parents about all these, including Sajjan's princely and elaborately decorated mansion.

While taking meals, Sajjan told Bijay about his late mother, whose picture was hanging on the wall. Bijay bowed down with folded hands from his dining chair. Though a napkin was given to Bijay along with spoons and forks, Bijay had never used them and felt extremely shy to use any of those. He silently watched as to how those were being used by his friend Sajjan and appreciated his not minding Bijay's etiquette and table manners.

Sajjan also told him about his younger sister, who was going to La Martinier, a very prestigious and posh school for girls in Kolkata. Bijay never asked Sajjan, but he voluntarily talked about his family known as 'Mica Kings' of India. They had several mica mines in Bihar, India. The family had several brothers, cousins etc., and they all owned their private houses in the big compound, where Sajjan lived. Every one of them had different offices, different businesses, and had ownership contracts on their individual interests. Bijay did not understand all these. It was difficult for him to visualize that all brothers and sisters, even though they were in one family, lived separately with their families, and rarely saw each other. Bijay wished that he could comprehend the meaning of all that Sajjan told him casually, never showing his financial strength, superiority and ego.

Realising that lunch was over long ago, and Bijay had to go back to his place, he indicated his readiness to leave. At that point, he surely wished that Sajjan's driver could drop him, as he saw several cars in front of Sajjan's house. Sajjan never made such an offer. Later, Bijay came to know that the driver was off, this being Sunday, and Sajjan did not drive yet. Sajjan probably forgot to show the rest of his home to Bijay. It did not matter to Bijay in any case. So, when he put on his shoes to go back, Sajjan came to the door and bid him off.

Bijay had to see the security guard, and, in his innocence, he asked him about the best way to return to his house in Burrabazar Area. The security guard offered to get him a taxi, but Bijay knew that it would be very expensive. So, he thanked him and decided to walk to the nearest tram depot in Ballygunj. He had to wait a lot before a tram car showed up. It was going only to Esplanade. He had to take another tram car from there to Burrabazar. He had to walk a lot for his house. Any way, he was happy to see his friend, who was so nice to him, despite all the affluence he saw today. He decided to tell all about it to his parents, who eagerly listened to him, and lauded him for knowing Sajjan and having contacts with him. In those days, contacts were everything for success!

Bijay did not see Sajjan in School, as it was off for summer vacation. Later, Bijay came to know that Sajjan had gone for about two weeks with his family, including his sister and father, to their summer resort in Nainital. Bijay had never heard about Nainital, and his family never took any vacation. His father used to go for work as usual, as it was essential for him to do so in order to meet his ends. When one of his father's close relatives, paid a visit one Sunday, he asked him to allow Bijay to join and help him in his cloth shop, not very far from their house. In the back of their minds, Bijay's parents thought it would mean Bijay would earn some money for his summer work and help the family in managing surmounting expenses. It did not turn out to be true, as Bijay was never paid despite his hard work every day for about a month. His relative felt that he was being trained and he should rather be happy getting an opportunity to learn rather than sitting at home doing nothing.

Time rolled on with nothing new happening to Sajjan and Bijay. Class after class, year after year, Bijay was, as usual, topping in his class. Sajjan was overall a very good student in his class, but he did not have that much luck in his school ranking. He was nevertheless happy for

Bijay and never had any complaint about himself. He became more engrossed in his family affairs and was usually in a hurry to go back home. So, he could not drop Bijay that often and Bijay also never asked him for a ride. However, whenever he dropped Bijay, he voluntarily offered to go up the stairs and say hello to Bijay's mom, who was always at home. She was always happy to see Sajjan, who was so simple despite all that she heard from Bijay regarding his riches and fame. She always offered some delicacies prepared by her in her spare time, but Sajjan insisted to have dinner with Bijay.

Sajjan really liked food prepared by Bijay's mom, and she herself enjoyed serving very simple dishes, and sitting in front of them, rolling a feathered mini fan (Pankha) for a little air. It was very warm inside the room, but Sajjan never showed any discomfort and uneasiness. It was so nice of him to take the stairs all the way to the fifth floor despite his being very bulky and heavy. Bijay learnt a lot from his friend-his simplicity and good manners without showing off any ego. It is surprising that in all these years, Sajjan was so relaxed and spoke freely with his friend Bijay, who was always hesitant to talk too much, as he always had inferiority complex about himself and his family.

Year of graduation from the High School was very near and Bijay and Sajjan both were studying hard to do well in the School Final Examination. Sajjan had already made plans to attend a very prestigious and expensive college, Saint Xavier's College, not very far from his house. Bijay did not have any plan, and frankly speaking, he left everything for his father to decide about him. He knew that the family was always short of money, and his parents would like for him to start working and earning some money for helping his father manage the mounting expenses. So Bijay stopped thinking about his plans and avoided the topic whenever Sajjan asked him about his plans after finishing School Final, equivalent to High School.

The results were announced in a couple of months. They appeared on school notice boards, local newspapers etc. Bijay was on the merit list and secured First Division with high Honors and Distinction. He topped not only in his Class, but also got a Government Scholarship of Rupees fifteen, equivalent to a quarter of a U.S. dollar per month, for college studies. Sajjan passed in Second Division, and he was happy with Bijay's and his performance. Both congratulated each other and invited each other to visit as often as possible. Sajjan knew that Bijay did not have any telephone in his house, so he made it a point not to request Bijay to call him at his phone. Both decided to keep in touch.

II.

SEPARATE PATHS

Bijay's life was bumpy year after year amidst uncertainties, scarcities etc., but he used to call Sajjan whenever he got a chance to call him from one of his relatives' places having a telephone. He saw Sajjan occasionally, and so did Sajjan. At no stage, Bijay asked Sajjan for a favor of giving him a job, and Sajjan also assumed that he was doing well.

Bijay went to a technical institution and was looking for a job, having obtained an Engineering degree. Sajjan, on the other hand, obtained a Commerce degree, and was helping his father in the family business. Sajjan got married in a very rich family, and as his marriage took place near Delhi, very far from Kolkata, Bijay never got invited, and he could not have attended it any way. It would have been very expensive for Bijay to go there and give some gift too! He too got married to a local girl, Manju and invited Sajjan to attend their marriage. Sajjan attended the marriage alone and decided not to bring his wife with him. He cheerfully attended Bijay's wedding and gave them nice gifts both to Bijay's newly-wed wife, Manju and to Bijay. He greeted Bijay's parents and met all his relatives. He was very nice indeed and was in a very relaxed mood.

Bijay tried his best to get a job in his profession, but he failed despite being a top graduate from his prestigious technical college. Even if he had approached Sajjan, his friend could not have utilized his skills. Bijay finally applied to some foreign universities and succeeded in getting acceptance for admission into a U.S. University. He also applied for some financial help, and in view of his top rank, he was awarded a Research Assistantship at the university.

This was indeed a big boon for Bijay, and he, with his parents' consent, decided to leave for higher studies in USA. His intention, at that time, was to complete a Graduate Degree (M.S.), and then return back home to support his aging parents. Before returning, he had the hope of applying to some local companies for a job. These companies would probably pay attention to his foreign degree. He did not realize that networking with top executives in companies in India was the only way to get a job. Unfortunately, neither his father, nor he himself, had any such contact with any such top executive.

As the U.S. opportunity of Research Assistantship was only for Bijay, he decided to call his wife later. Obviously, arranging money for her tickets etc. was the main hurdle.

Before proceeding fpr USA, Bijay had to obtain a Student Visa from the U.S. Consulate at Kolkata. There were tons of formalities to be completed before his application for visa could be considered. He had to get an Indian passport from the local office at Kolkata. To his dismay, the passport application form required notarized photographs, address verification by the police, income tax clearance (though Bijay did not have any income) from the tax office and a big amount as application fee.

Bijay's father encouraged him and provided him financial support. As the deadline for Registration was very near, Bijay was not left with any time to inform his relatives and friends including Sajjan. Once he got passport, running here and there, to complete other requirements, he approached The U.S. Consulate for a Student Visa (I-20). The latter was a must for him to enter USA, secure admission for Graduate studies, and work 20 hours a week as a Research Assistant at the University of California at their Berkeley Campus.

It was indeed a unique experience for Bijay, who was given a very lengthy visa application form detailing all requirements precisely. He was clearly told at the Consulate counter that everything must be completed systematically as per the requirements, and if Bijay failed in fulfilling any requirement, he would be rejected for a visa, and could not register in time at least for a year. Bijay was completely lost, but he had no other alternative. He had the support of his parents, who were hard pressed to keep on paying Bijay day after day without any guarantee that he would succeed in getting a visa in time.

Visa application required a valid passport, all educational certificates, vaccination proof, notarized photographs, very elaborate medical examinations including x-rays etc., financial capability to support himself for a year, air ticket to USA etc., etc. Bijay somehow completed all these formalities working day and night and was finally able to get the Visa from the Consulate. This did not allow him any time even to contact any of his close friends, including Sajjan.

When Sajjan came to know about Bijay's sudden departure to USA, he sent a letter of good wishes to Bijay. He got his U.S. address from his parents. He affectionately wrote to Bijay that he wished he could have informed him, and in that case, he would have certainly seen him off at the Airport.

The Student Visa enabled Bijay to study only, and not work in USA except for his Research Assistantship. Bijay had to work very hard in the new environment almost day and night. He wanted to show his worth to his Graduate Adviser. He successfully finished his graduate work for Master's degree in Engineering in about a year, and his research work was highly appreciated by his Graduate Adviser.

Bijay's Advisor and he himself both finally published a joint paper based on Bijay's work, under his guidance, in a reputed technical journal. Unfortunately, his Advisor could not support Bijay for doctoral degree under him due to lack of funds. He, however, recommended him for a job in USA. Bijay knew that he had only a student visa and could not work unless he applied and obtained a work visa. As technical people in his profession were in great demand at that time in USA, Bijay not only got a nice job in his profession, but also obtained a work permit with his employer's help.

Bijay worked very enthusiastically, and his employer was very happy. He missed his family back home and with his employer's help, he got his wife, Manju, and two sons, Vikash and Vishal, to USA. He knew the work situation there in India and accepted the U.S. job gladly without any hesitation. He was so engrossed in his work; he did not keep in touch with friends and relatives back home. Time passed very fast, and his parents were getting old. They wanted to see him back home at least once before it was too late!

Bijay needed at least 6 weeks to travel that far, and see his parents, friends and relatives. His company granted him time off, as he was accumulating his vacation, as per company's rules. As it was extremely expensive to travel that far, he went alone. His long flight was quite smooth, and on reaching Kolkata, where his family lived, he passed through Sajjan's house, and memory took him several years back. Luckily, he still had his telephone number, and decided to call him during his 6 weeks' stay.

III.

REESTABLISHING CONTACTS

When Bijay called Sajjan, nobody picked up the phone, and Bijay wondered as to whether it was the right number. May be the phone number was changed, as about ten years had passed since his friend Sajjan gave him that number. Bijay decided to call him again on checking the telephone directory, but there was nobody by that name in it. He called the telephone directory office and was told Sajjan's name was not listed there. Bijay got a bit concerned about Sajjan and decided to go all the way to his house. He still remembered his last trip to his house several years back and did not forget that he had to pass through the security guard at the massive gate. The Security Guard might turn him down as he did not make any appointment with Sajjan or his family. He knew Sajjan got married long time back, and it was not at all right to go without appointment or invitation. He kept on struggling in his mind, and finally took a taxi, as he could afford one now.

The taxi took Bijay to the right address and Bijay, in spite of so much time lapse, recognized the place, where Sajjan and his family lived.

This time, the gate was securely locked, and there was no guard either. Who could help Bijay? Desperately, Bijay decided to ask one of the Panwallas (betel leaf maker) near his house. He did not know anything about the big mansion complex, Sajjan etc. The shopkeeper, as a courtesy to Bijay, asked one of the old 'Jhakawallas' (merchandise carrier) as to whether he knew anything. His answer was plain 'no'.

Apparently, it was a futile trip to his one-time closest friend Sajjan's house. Next day, he told his father about his inability to contact Sajjan with a request for his help, as his office in Jute Brokerage House was not too far from Sajjan's father's office. Even with repeated telephone calls made by Bijay's father from India Exchange, nobody responded from Sajjan's father's office. This aroused great deal of concern and anxiety in Bijay's mind, and he blamed himself for keeping quiet from Sajjan for such a long time. Time was running out, as Bijay had to return to his job in USA. Everything else, other than desperate attempt to contact Sajjan was going alright, and Bijay could see all relatives one by one. It refreshed contacts with all his and his wife's relatives, and close friends except Sajjan. Bijay had a heavy heart and only wished all the best for his close friend Sajjan and his family.

IV.

FRUSTRATIONS

Bijay left for USA as per his planned schedule. He did not forget his unfinished task of contacting Sajjan. When he got some time, he dropped a letter for him by air mail at his address, but it was returned by post office by surface mail with the remark 'incorrect address'. He was very disappointed. While at Kolkata, he had called the local telephone directory office, and asked for his new telephone number. There was nobody by that name and address other than the one that Bijay had. How could the things change so suddenly and abruptly? Bijay had contacted St. Xavier's College, where Sajjan had his Commerce Degree, but they did not have any address other than the one that Bijay had.

Bijay distinctly remembered the prosperity and affluence of Sajjan's father. He also remembered that Sajjan himself got married in a very prestigious and rich family near Delhi. All his cousin brothers and uncles had individual posh houses in the huge palatial complex that he had seen under heavy security. Sajjan had told him that all of them had their own business ventures and managed them individually. How could such a solid empire change? May be they all moved to a new place, but there was no forwarding address.

Finally, though frustrated, Bijay decided to put the matter of locating Sajjan off for the time being till he visited Kolkata again. Bijay had a very demanding job with strict deadlines to be met in order to survive. He could not overlook the requirements of his job- no matter what was making him concerned about his closest friend Sajjan and his family. Obviously, job was the first and topmost priority with Bijay.

In his spare time, when Bijay thought of finding his missing friend, he had several ideas. He realized fully well that he was in USA, and he was trying to find out something in one of the most populous cities of India. He also knew that even back home in India, nobody had time to care about Sajjan, even if he requested somebody. Nevertheless, he took the courage to write to his father to get some information about Sajjan, his family, closed gates etc. Sajjan's house was so far from his father's rented residence that it was not possible for him to spare so much time to go to Sajjan's address except during one of the holidays.

When Bijay's father went there, he also saw the gates securely locked without any guard outside the walls of this big compound and en-quired several people, including some shopkeepers, but nobody knew about them. In fact, many of them got skeptical of him enquiring whereabouts about one of the richest and most influential families of Kolkata. He wrote back to Bijay about his returning back without any success. Before giving up, he talked to several of his fellow workers and friends as to whether they knew about the big financial business and their offices. He himself walked with them only to find a multi-storied air-conditioned building completely locked without any security. He, somehow, came across a police officer and asked him, but got no information except advice to go to the police station and lodge an enquiry.

One information or the other added to the complexities in Bijay's mind. He did not wish to think any more about finding Sajjan's whereabouts, but he could not help himself. He had such a nice time with him and regretted again and again that he did not keep in touch with him regularly. Sajjan was nice to Bijay right from day one till he saw him last. When Bijay left for USA without informing him, he wrote a very nice letter to him at his U.S. address, congratulating him for going to USA and wished he had known about it, so that he could at least see him off at the local Airport. This reflected Sajjan's nice feelings for Bijay.

Bijay knew it would be several years before he would go to Kolkata to visit his parents back home. This was the time when he got such an opportunity to go alone, as he could not afford air tickets for his entire family in USA. His memory flashed back when he passed through his friend's house and the quest started without any real reason other than his friend being so nice to him without asking anything in return from Bijay.

Bijay and his family, consisting of his wife and two sons, were doing well in USA. All of them were very busy. Really speaking, family affairs, full time job, attending college in the evening for an Engineering Management Degree did not leave much time for Bijay to think about anybody and anything else. He was accumulating his vacation as per the Company rules, anticipating that he would eventually be able to go to Kolkata all by himself, and devote more time finding about his one-time close friend, Sajjan. Then he reconciled himself in his mind that even if he would be successful in finding his whereabouts, what would he do?

It was beyond Bijay's imagination that he could help Sajjan in any way. He sincerely wished that there was no health, major problem, such as accident etc. He ruled out this idea in view of the fact that the entire complex, where his friend Sajjan and his family lived, was deserted and locked. How could such a major thing happen without anybody outside the complex knowing about it? Was it due to a major robbery, flood or fire in the complex? Did they all decide to sell the Complex, and move to another?

Finally in a couple of years, he told his wife about his intention to go to Kolkata. She knew very well the unnecessary tension Bijay had in his mind about Sajjan and his family. She herself wanted to go with him, but her work did not allow her to take off for such a long time (about 4 weeks). Then what will happen to the children? Their schools were still on and a lot of attention including taking and bringing them to and from the schools were parents' responsibility. Apart from schools, both participated in soccer, tennis etc. At least one of the parents had to stay back. So it cleared the way for Bijay to travel alone again to Kolkata. Truly speaking, air tickets were so expensive that he could not afford to take all of them with him.

On arrival at Kolkata International Airport (Dumdum Airport) in the middle of the night, almost exhausted and half-sleeping, Bijay once again passed through Sajjan's house. He did not stop the taxi, as it was very dark, and he had already planned to put a lot of time for locating his friend's whereabouts. It took him several days to get over the jet leg (time gap) of about twenty hours' air travel from USA to India. Fortunately, he was with his parents and rest of the family, who were very happy to see him again. This time also, Bijay brought some nice gifts for each family member, including his wife's family in the same city. He saw them also once he was back to normal. Fortunately, all of them were doing well and only wished to see Bijay's entire family soon. Bijay assured them to do his best.

In about a week after his arrival, Bijay got dressed as usual, and left for Sajjan's house. He knew that he had a lot of things to complete before returning back to USA, including doing some shopping for his wife and family in USA, reconciling minor bank accounts in India, seeing his close relatives, friends etc. US tax laws required reporting foreign balance, interest etc. Bijay did receive interest and dividends in India, but no statement was ever received. In order to save time for these errands, he took a taxi to Sajjan's house. He got down from the taxi, but there was no security guard at the locked door, and there was no notice/ information outside the gate.

Bijay saw some hawkers outside the complex. He took the courage to ask them about the residents of the big Complex. All that he was told that they knew nothing about them. From whatever they said, the only conclusion was that the Complex was deserted long time back, may be several years back. He walked outside the big wall of the compound, hoping he would find some clue about Sajjan and his family. He saw some apparently established and long-time shops outside the Complex, but none of them knew anything about them. He could not see any catastrophe like fire, flood etc., as the buildings were still there, and they still looked impressive even from a distance.

Bijay suddenly remembered his father coming here long back and he wrote to Bijay in USA about talking to a policeman who advised him to go to the local police station and lodge an enquiry. Bijay did see a policeman in uniform not too far from the complex. He walked to him and greeted him. The policeman rudely looked at Bijay with a very skeptical reaction. More than usual, the policeman is contacted by a pedestrian like Bijay only when there is something related to law, crime and accident. Bijay politely asked him about the residents of the big complex, pointing towards it.

The policeman rudely asked him about Bijay's connection regarding the complex and its residents, and about his intention of knowing about them now. Bijay, in a very mild tone, told him about his one-time close friend Sajjan living there. The policeman grew more suspicious about Bijay, who was walking on the road, had no car and was in a very simple dress. It was very difficult for anybody to believe Bijay's knowing Sajjan or anybody in that palatial complex!

Though the policeman shrugged Bijay rudely, he mustered strength to ask nearby general merchandise shopkeeper about a nearby police station. He pointed from his shop that it was not too far and also gave direction to Bijay to walk over there. For the first time in his life so far, Bijay went to a Police Station-a red building saying Ballygunj Police Station. He was at least able to find it and he had a faint hope of getting some help there about his problem. So he walked inside and he found a big crowd. There were so many floors in that building and it was very difficult for Bijay to go to the right floor and Office (room) to get the type of help that he wanted. He enquired from some passers-by, and everybody was confused about Bijay's problem! Though the City Of Kolkata is one of the most

crowded cities of the world, Bijay's types of cases in the police sta-
tions were very rare.

When Bijay did not get any satisfactory reply, he decided to stand in
one of the less crowded queues. By any standard, he figured out that
it would take him the entire day to come to the window, realizing the
number of people ahead of him. In some cases, he noticed that it took
a lot of time, in some others,less! He thought it must be because of
the type of problem. He almost came to conclusion that he might not
even make it to the window that day within working hours of the police station to lodge an enquiry about his friend Sajjan, his family and
the Complex.

Bijay talked to a man standing in front of him and was told that some
people stood in line from early morning even before the police station
opened at 8 AM for business. Many are turned down at the closing
time of 5 PM sharp in the evening, however urgent the matter is! No
work was done in the lunch hour between 12 noon to 1 PM. He was
also told that most people were there only because they were summoned to appear in the police station that day. In fact, the person
standing in front of him was surprised that Bijay did not have any
summon or police case against him. He was smilingly wondering
about Bijay's standing there for hours in the big crowd waiting for
his turn to come!

Bijay was right about his speculation that he might not be near the
point man till the closing hour. If he still wanted to pursue the matter,
he had to show up there early in the morning before sunrise and stand
in queue to take his turn for lodging his peculiar enquiry about something, which theoretically should not be of any concern to him, much

less to the police station, which had more complex cases like crime, accidents, thefts etc.

Thoroughly disappointed, he returned home taking a taxi, which was very expensive in the rush hours. He was very tired and was very glad to be back home. When his mother enquired about his day, he did not reply and politely told her to let him take some rest. She could see that he was bothered about some problem, but she realized that she could not help him anyway. Bijay relaxed for a while and spent rest of the evening and night with his family members. He did not want to express his disappointment and total failure, realizing fully well that every day was making his short vacation shorter and shorter! Before he turned to bed, he planned to take another shot about his wild venture next day right from early morning!

Bijay's parents were already up, as usual, very early in the morning. Bijay also got up early and told his parents that he was taking a morning walk and would be a little late. He also told them not to wait for him for lunch, but he would surely be back by dinner time. He took a taxi and went straight to the police station. It was about 6 AM. At that time, the police station building was still closed, but people were already lining up very fast, though the crowd was much smaller.

Bijay did not know the proper floor or window, so he just stood in one of the smallest queues. At about lunch time, he was very near the police officer at the window, and was hoping to see him. Right at that time, he was told that it was lunch time, and the window would open only in an hour. Bijay could not leave the queue, as he knew that once he lost his turn, he would lose his spot and had to start in a very long queue all over again. He just could not afford to do it, as he had already lost one and half days over this rather stupid and unrealistic problem. He was not convinced himself that he was not wasting so much of his rather busy time over this worthless issue, which really did not concern him!

In an hour, the window opened, and Bijay's turn came just an hour before quitting time. So, he nervously explained his problem to the police officer, who appeared to be in a big hurry, and did not see any point in talking to Bijay without any solid proof or evidence on his missing friend. He, however, gave him an official form to lodge his complaint on another floor in a particular office. He must complete the form with all other requirements before Bijay could turn the form for further consideration. He must go through a big line of queue on completing this form once again, if he wished to pursue the matter further.

Bijay was extremely tired and disappointed, and returned home. He felt relaxed after a hot cup of tea prepared by his mother. He did not have any courage to explain to her the reason for his exhaustion. He took a nap and felt a bit relaxed. He had only two choices, either to drop the idea of locating his friend or complete every formality on the form required by the police officer. The idea of dropping appealed to him first, as this would have meant utilizing rest of his time at home in vacation and relaxed mood.

When Bijay remembered the second option, he realized it was, in fact, the primary reason of his coming for vacation without family, so that he could use his entire time to achieve his objective- the intense desire and acute concern for his friend Sajjan. He could not sleep the whole night pondering about his priorities. Finally, he decided not to give up in his efforts, realizing fully well that he would never be able to put this much time in future for this purpose, and it might be terribly late by then.

In the morning, he looked at the lengthy form with some questions about Sajjan. He could not possibly answer them. For example, he did not know his friend's date of birth, his parents' correct names, the length of exact time they stayed in the magnificent complex etc. He could not even remember the name of his sister. Suddenly, a thought came to his mind that Sajjan mentioned about his sister going to La Martinier, a prestigious and posh school near their complex. Though Bijay had never seen her, nor the school, nor did he know its exact location, he decided to go there by a taxi and enquire as to whether her sister had already graduated. Possibly, then, he could know about the family's whereabouts etc. It is strange that when a man faces nothing but disappointment, even a flash of an idea appears as a ray of hope for him.

When Bijay reached La Martinier, he saw a security guard at the gate. Bijay asked him as to whether it was possible to talk to somebody in the office about his friend's sister. The security guard took his name and asked him to stand outside the gate. To his pleasant surprise, he was allowed to go inside and see one of the lady clerks in the office. Things looked a lot more systematic and organized than those in the police station. Furthermore, he did not have to face any crowd or stand in line to take his turn. Bijay explained the reason for his going there, and wondered as to whether he could get some information as a favor from the lady clerk. The lady clerk wanted to know the name of the girl and unfortunately, Bijay did not remember her name at all. He, however, gave the address and telephone number correctly, as given by his friend Sajjan long back. Luckily, he saved them, and noted in the address book Bijay had prepared long back before leaving for USA.

The lady was finally able to, after going through a number of old registers, locate that address and phone number. Bijay was told that her name was Manisha, who checked out of the school about seven years back. She did not leave any forwarding address, and there was no way to find out about the family's whereabouts now. The school did not know anything about the big mansion locked now, without anybody living there and without any security guard. So, Bijay was again at the starting point and he did not achieve much. All that he knew that something happened about seven years back, three years after Bijay's departure to USA, and Sajjan's sister's name was Manisha.

In the morning, Bijay on finishing his breakfast, decided to visit his relatives, rather than worrying about his friend. Wherever Bijay went, he could not help his mind, repeatedly turning back to his main purpose of this very expensive trip for four weeks, viz. locating the whereabouts of his close friend, Sajjan. When he went to see his maternal uncle with his mother, who appreciated Bijay's gesture in taking her in a taxi, which she could not generally afford without any solid reason. She also used this opportunity to do some shopping with Bijay.

Bijay suddenly realized that he was in Burrabazar Area of Kolkata. In fact, his old school, where he studied till High School with Sajjan, was very close. Once he dropped his mother to her cousin brother's place, he took everybody's permission to visit his old school, and return in a couple of hours. It was alright for everybody, as it was too early yet to have lunch, which was not ready. As Bijay's mother visited her brother only occasionally, lunch had to have some special dishes, as per family's tradition.

Bijay walked to the school, which was very close. He remembered the narrow and crowded streets that he used to cross every day twice, while coming to the school, and then going back from the school and returning home. Nothing had changed. In fact, the streets were more crowded, and fully jammed with hardly any room for pedestrians to walk and cross. Somehow, he made it to the school. He remembered the nice time that he spent in the school. He went straight to the Headmaster's room, and was told that practically all teachers, including the Headmaster, had retired, and there was new staff. He waited for the new Headmaster for some time in the hope that he might be helpful to him and wanted to come straight to the point about what he was looking for (Sajjan) !

When the Headmaster arrived there, Bijay respectfully came straight to. the point, and explained the purpose of his visit. As he was an ex-student of the school, he got maximum attention. The headmaster, with the help of an Admissions Clerk, located the yearly Registers, one by one, from about 17-20 years ago. It was quite a bit of work. Two hours were already gone, and they were still searching.

Everything in those days was written by hand, as there was no computer, no internet, no e-mail, no disk etc. As they were almost at a breaking point, one clerk finally located a Register for the year 1955 (now it was 1972), and found both Bijay's and Sajjan's records. Bijay smartly noted Sajjan's birth date, place of birth, his parents' name and his address at that time. Though there was a record for Bijay going to Presidency College for I.Sc., Engineering Degree at ISM, record for

Sajjan mentioned his going to St.Xaviers's College only. Nothing was available after that.

Bijay had taken a lot of time from School authorities, and he sincerely thanked them and later on acknowledged their valuable time by a letter from USA. School wished to see him again before he would leave for USA. They wanted to have a reception for him in the school, and wanted to hear about his experiences and life in USA. Bijay earned a big name for the School and impressed all authorities, teachers and students. Bijay assured them that he would try to come again with some advance notice to them. He took Headmaster's telephone number, though he knew he was very tight of time mainly because of the complex puzzle he was trying to solve!

Bijay was rather happy for getting some critical information about his friend Sajjan from the school, which both attended several years back. This was in fact the best day for him, and he returned to his maternal uncle's home a bit late, where his mother was anxiously waiting for him. They both had a sumptuous lunch. Seeing Bijay happy for the first time since he came about a week ago made his mother happy too. In the late afternoon, Bijay took her whichever shop she wanted to go and bought her whatever she needed. It was a rare opportunity for Bijay to be of some help to her in several years, and she always used to thank her son for this courtesy till she lived. Bijay did so little, yet she was so happy! He never realized that mothers are always like that throughout the world, and they all are the rarest, most selfless species filled with love and only love for their children, no matter what!

When Bijay returned home with his mother by a taxi in the evening, he looked at the multi-paged form given by the police station. Today's accomplishments made him feel that he could start filling it up, though there were still so many unknowns. Suddenly, he became so strong that he boldly wanted to see the police station again in a few days, even though the form had many blanks. In the worst scenario, he would be turned down, but it was the best he could do in those circumstances.

Bijay distinctly remembered that Sajjan came alone to his marriage, and chose not to bring his wife with him, otherwise he would have probably introduced his wife to Bijay. Unfortunately, Bijay did not remember the day of Sajjan's marriage, the place (which was far from Kolkata and near New Delhi) and his wife's name. Some photographs were taken on the occasion of Bijay's marriage, and he quickly looked at the album. He could find only Sajjan's photograph standing with him.

The police officer required the missing person's identity, and Bijay thought that besides giving his wedding day photograph with Sajjan, he would mention in the form that he was bulky at that time. Rest of the week, Bijay relaxed at home, and put a few minutes each day filling up the form as precisely and meticulously as he could without much information about Sajjan. In his school days, it never occurred to him to find out more about Sajjan. They were close friends by all standards, and he was very nice to Bijay. He wished he could keep in touch with him, so that he did not have to face all these hurdles in locating Sajjan. He just wanted to give him his best wishes, greetings, and possibly a small gift that he got from USA for him, hoping to find and see him during his stay. Though Sajjan never expected anything from Bijay, the latter thought he would fill up the gap this time.

By the time, Bijay was ready to complete and hand over the form, it was already the last week of his stay, and he still had a lot of pending things that he must complete before leaving. Nobody knows as to when he would get a chance to return to Kolkata, even though he would like to return soon to follow up his quest, and see his aging parents. Next time, he would bring his whole family, as his wife's parents were also getting old. Both sides wanted to see their grandchildren, as it would indeed be a long gap since they saw them last time before their leaving for USA.

Bijay worked hard and completed the lengthy form. He took it along with the only photograph that he had with his apparently missing friend, Sajjan, at the time of Bijay's wedding. He had to stand in a big line again and finally after four hours' standing in line, he was successful in seeing a police officer at the window. He handed over the form to him. The police officer was very, very annoyed to see the incomplete form and handed it back to Bijay to fill it up completely and then only come back again. Bijay, this time, frankly, told him that he would not be able to find any more information, as he had already exhausted all the possibilities. Besides, he also told him for the first time that he would be leaving for USA by the end of the week, and it would possibly be several years before he would return.

Bijay very humbly requested him to keep the form and the photograph. He told him that even if the required form is rejected due to incomplete and insufficient information, probably putting it in waste basket, it was alright for Bijay to leave it there, as he would fully understand the situation. Bijay's words were so pathetic, and his face was so saddened that the police officer asked him to wait outside the big queue. The police officer told Bijay that he would talk to his boss as a special case, and he might approve it for further action in due course. He made it crystal clear that such a complicated case might require a lot of time, due to several investigations to be carried out. It might not be a case of all residents in the complex moving elsewhere. The police officer advised Bijay to tell his boss about his leaving for USA and not returning soon, as that would draw his attention seriously.

Bijay agreed to wait outside. It was almost the closing time for the police station. Bijay thought that the busy policeman might have forgotten about his case and might not come back to him at least that day. Right at that time, another elderly police officer called Bijay by his name, and asked him to follow him to his office. It was a very impressive office indeed, and from whatever Bijay saw in his room, it did not take much time for Bijay to realize that he must be a very senior police officer. He was very seriously looking at the form and the photograph, given by Bijay earlier. Finally, he asked Bijay to note down his US address on that form and advised him not to expect too much, as the form did not have complete information. He assured Bijay that he would forward it for further action in due course.

Bijay thanked him sincerely, as it was indeed the first time, he was able to accomplish something, even though he knew that anything was possible, from nothing to something, because of his incomplete information on the form about his missing friend, Sajjan.

Bijay left for USA as per schedule, as he had to go back to work, and did not have any more vacation. Besides, his family was there anxiously waiting for him to return. Every time, he called his wife, she wanted him to try to return earlier. He must attend to some urgent household matters on return, as it was already a long vacation. She warned him that as some payments were long overdue, Bijay would have to pay finance charges.

Bijay returned to his old job in USA. As usual, day after day, his job, family chores, evening school etc. kept him tied up. Soon, he got himself so much engrossed in his activities there that he almost forgot about his trip, home and his fruitless efforts to locate his close friend, Sajjan. He did not tell all about these to anybody, including his wife, as he, somehow, wanted to prove that he enjoyed his month-long vacation, which was indeed very expensive and exhaustive. He himself wanted to forget about the ordeal about missing Sajjan, as he had done whatever he could, and now it was up to the police officers in Ballygunj to inform him if they could locate the whereabouts of Sajjan and his family.

Almost two years had passed. Quite unexpectedly, his father in Kolkata informed him in USA that there was a letter from the Ballygunj Police Station for Bijay. He did not open it, and therefore did not know the contents. Bijay usually contacted him by air-mail letters, which took about 2 weeks to reach him. This time, Bijay got so curious about the unread letter that he called him long-distance. His parents were very happy to hear his voice, as it was already a very long time since he talked to them. Their health was in bad shape due to old age, and they were feeling very lonely all by themselves without other family members near them.

Bijay was very curious about the letter from the Police Station, and he requested his father to open and read it. The letter disappointed Bijay more, as first, it said that they could not verify that the missing person was ever staying at the place, mentioned by Bijay in his form. Also, they could not find anybody around there, who ever heard about the family living there in the big Complex. It was unimaginable for Bijay to think the big family never had occasion to mix with anybody outside the

Complex. When he thought hard, he consoled himself by realizing that probably all their requirements were met by the guards, drivers, servants, helpers etc. Maybe they wishfully wanted to avoid company outside due to security reasons.

Bijay thanked his father for informing him about the long-awaited letter and now reading it over the phone, though his father was not too happy about his very busy son worrying about a forgotten era and lost contact. So, Bijay was at ground zero, and did not know any other way to locate his close friend Sajjan. Suddenly, he remembered about Sajjan's father's office near his father's office long back. He kept pondering about it deciding as to whether it was worth the trouble for his ailing father to go to the office area in India Exchange and try to find out about his father's office. Bijay very well was aware that Sajjan's father would never talk to his father and give information about Sajjan. Even if Sajjan was there in office, he would never allow him to talk to a stranger, specially as it was completely a non-business visit by him.

Restless as Bijay was, due to obvious reasons, he decided to request his father to help him on this matter. Though his father had not gone to his office in years, as he was retired, and too weak to walk in the very dangerous, extremely crowded Business area, he agreed to do something per Bijay's strange request. Before going there, he was almost sure that he won't be able to find anything worthwhile, as he won't even be allowed to go inside the building, nor did he know anybody there. Everybody was so busy there, and nobody wanted to talk to a stranger like his father asking something, which appeared to be irrelevant to them.

Bijay's father, in spite of weakness, helplessly, took a taxi and went alone to his old office, just for the sake of his son, Bijay. Everybody greeted him, as it was nice for them to see him after such a long time. Everybody greeted, as he was always nice to each and every person in the joint office. When asked about his purpose of going there, he quietly avoided the topic. After a little bit of rest, he walked to the place that he knew was probably the Office Building of Sajjan's father. He had to face a lot of crowds, many times almost losing his balance, but he did not give up, as it was his son Bijay's request! When he reached there, he faced two fully armed security guards at the entrance. He politely told them about the purpose of his going there and requested them to help him as an old and extremely weak person. They had never seen him in the building. Bijay's father was thoroughly disappointed and asked them if he could go inside and talk to somebody working in one of the offices for a long time, about 20 years back!

It was indeed a wild shot. One security guard took him to a very old and experienced person, who on hearing the story from the guard briefly, realized that Bijay's father was too weak and tired. Though he appeared to be extremely busy, he offered him a chair and asked him to sit down and wait till he could spare some time. He offered him a glass of cold water, which was indeed a blessing for Bijay's father.

Fortunately, Bijay's father was very fluent in Bengali language and thanked him profoundly for his kind gesture. When he was available, his first question to him in Bengali was about the reason for seeing him today. On hearing the entire matter, he regretfully explained his not knowing anything about the family business of Sajjan, his father and the Agarwal family. He also told Bijay's father that, to the best of his knowledge, the entire office space in that building belonged to the Income Tax Office, where he had been working all his life. Though it was a complete shock to Bijay's father, who wondered as to whether he had come to the right building. He sincerely thanked the old officer, who spared so much of his busy time for him.

On coming out of that office building, Bijay's father thought of briefly glancing at more office buildings, one on each side, but those belonged to some British Companies doing business in India for a very long time. He finally gave up as it was late in the evening, and returned home by a taxi, which was very, very difficult to get during the busy business hours! He had no choice but keep trying for a taxi, as he was too weak to go to the bus stop or tram depot. The taximan wanted him to share with 4 other passengers, and yet pay full fare! He had no choice but to agree.

On returning home, he told Bijay's mom, who was extremely worried, about his futile trip to India Exchange to help Bijay per his request. She advised her husband to forget about it, even though she had a good impression about Sajjan and remembered his good manners, in spite of being such a rich person!

Bijay in USA was anxiously waiting for his father's findings. His father promptly wrote a letter to him and explained his unsuccessful attempt to locate his friend's and his father's office. He tried his best, but he was sorry that he could not find anything that could help Bijay. Part of the reason was that it was such a long time and considering the extremely busy business area and constantly changing environment there, it was almost foolish on his part to even entertain Bijay's request and go there. Anyway, now that he took so much trouble, particularly as he was very weak, he indicated in his letter his complete helplessness.

Bijay, as usual, got the air-mail letter in 15 days and was disappointed, but not surprised, to get the news of his father's fruitless effort in locating his close friend Sajjan. He felt very sad about it for 2-3 days, then he calmed down, till he gradually forgot about it. Days after days, months after months, a couple of years went by. It almost came to a point where Bijay did not appear to be bothered about this anymore.

About 3 years had gone by, and Bijay decided to have some vacation with the whole family. He and his wife, Manju, both carefully saved and accumulated their entitled vacation time, as per their respective company's rules. They had 2 sons, Vikash and Vishal, who were very bright students in their respective schools. Both were having some time off (holidays) due to Thanksgiving, Christmas and New Year Celebrations throughout USA. Air tickets at this peak vacation time, when weather in India is best, cost Bijay a lot. The family travelled together, and again, Bijay passed through his close friend's house almost in the middle of the night. He could not see much as it was dark, but his memory was rekindled with happy moments he had spent with Sajjan, whom he and his father could not contact in spite of all possible efforts, including reporting to the local police station in neighborhood. Nothing came out of all their efforts, and it was still a puzzle to be solved.

Bijay was glad for his trip home, though it was very expensive indeed! He could see his aging parents and relatives. Likewise, Manju, his wife, could see her old parents and relatives. The children were enjoying, as weather was just perfect for them, as it was winter- not too cold, not at all hot. The city of Kolkata was decorated nicely with lights all over and all local shops were having special holiday sale.

Bijay was really relaxing this time, as he dropped the idea of pursuing Sajjan's matter, even though he often thought about him since he passed through his palatial house in the big complex. He wondered as to whether it was still locked without any security guard at the massive entrance gate Bijay remembered that it had to be opened by rolling it, and the security guard made it a point to close it immediately on allowing the cars inside. There were hardly pedestrians like Bijay, who came on foot or by a taxi outside (not allowed in the complex). In such a situation, the guard used to allow the visitor or relatives to enter through a small window, which was only rarely opened! The guard must be informed in advance by the concerned complex resident to allow such a person, who was checked thoroughly.

One Sunday, when Bijay was having tea with his father and reading local Sunday newspaper, he was wonder struck to see a notice in the corner of one of the back pages of the paper. It said that the income tax department was auctioning the big complex, where his friend was living with his cousin brothers and their families. He wanted to be sure that the complex address was the same that he got from Sajjan long back, and he promptly checked and verified it. Why did the tax department come in picture? Why after such a long time? He was so curious to know about the whole matter that he took the courage of asking his father about it. He simply could not figure out anything by himself, and though hesitating, he politely asked his father. As his father was at home, and was in no hurry, partly as he was retired, and partly as he was relaxed, and felt good having tea with Bijay in a long while.

Bijay showed him the auction notice by the tax department. His father, based on his experience in business matters, tried to explain to Bijay that it could be that the family did not or could not pay the income taxes, which were due from them. He, however, made it clear that the amount was probably too big, and maybe they did not agree with tax audits. He also guessed that such a powerful and financially capable family must have challenged the tax authorities for years and years.

May be when tax department could not resolve the matter legally, they must have given them enough time to pay, if not in cash or check, may be in kind! Finally, it could also come to a point when the tax department was forced to evacuate them and forcibly take possession of the complex. He went beyond it in his guess work and pointed that possibly the tax payments were so huge that tax department seized their businesses and homes, forcibly after giving them enough notice, and taking all legal litigations against the family.

Bijay suddenly remembered Sajjan mentioning about his cousin brothers having completely different business involvements in various places. He could not, however, imagine and he came to know eventually that the building was seized by the Income Tax Department due to a huge tax debt accumulated with interest and massive penalty over the years.

Knowing Sajjan, Bijay had absolutely no doubt that he and his father must have paid their dues in time. There was no way that he could even imagine that anything like tax evasion would have occurred in Sajjan's business, as he must have joined his father. Was there any way to investigate further on the part of Bijay, so that he could locate his missing friend? Bijay's mind was unnecessarily circling around the Auction Notice by the tax department in possibly all local newspapers. He boldly asked his father as to whether there was any point in contacting one of the tax officers responsible for collecting taxes in that area.

Honestly, Bijay did not know anything about taxes, and he was completely nervous about enquiring further. The only thing, he knew, from his visit last time that he filed an enquiry request to the Area Police Station. They had already rejected his request on the ground that nobody was living in the Complex by the name of Sajjan, and particulars mentioned by Bijay in his form could not be verified. At least, they did not have any record about the family in their files. Even by detailed enquiry, they could not find anybody living near Complex, who even knew anything about Sajjan, his father, his family and his cousin brothers' families, what to say about their businesses and dealings!

Now that Bijay had some evidence on the big complex on auction by the tax authorities, he wondered as to whether he should take this information to the local police station. He smartly took out the cutting on the auction notice in the local paper. He also carefully found out the copy of the enquiry request he made during his last trip about 3 years ago. He also found the only reply he got from the police station about 2 years ago.

Bijay knew it would mean wasting some time on an issue that normally should not concern or bother him. He knew that his vacation was too short, and he had to do a lot before returning to USA. His mother was annoyed last time, and she would be upset by the unnecessary concern on part of Bijay. His father, though Bijay did not consult him on his idea, was probably equally upset by the auction notice on such a big business, that he heard about, from one of his friends long times ago. His friend unfortunately passed away several years ago and since his retirement, Bijay's father did not hear even a word about them.

On Sunday, when Bijay was having tea with his father, Bijay took the courage to ask him about the idea of approaching the local police station again with their letter and the auction notice on the big complex. Luckily, his father had kept the letter from the police station safely, and his father agreed that they would at least recognize that letter and try to listen to Bijay. Having got some encouragement from his father, Bijay decided to go to the Ballygunj Police Station. He told his mother, wife and the whole family that it would probably take the whole day. His father was too weak to go with him, and after all, what could he do?

Bijay took a taxi and proceeded to his destination. Nothing had changed since he came here last time. There was a very big crowd of people, but they were all systematically standing in queue for their turn. Luckily, Bijay remembered the exact place, where he submitted the required form. He had a copy of that form with him. He saw a much bigger line and he realized that it would take a long time before he could reach the window and talk to the Officer. In the worst case, he would not hit the target till the closing time, and he regretted that he did not come here early in the morning.

Bijay really could not afford to come here again, as he was almost near his departure date as per schedule. His wife was counting on him to do his shopping with her. She knew that she could depend on him for other musts like getting latest bank statements, confirming their bookings for returning to USA etc. They all knew that there was a huge penalty for making any change in the airline bookings. Besides, it was a rush time for travel, and it was almost impossible to get any vacant seat. They were, in fact, lucky enough to get reservations in this peak holiday period, only because of an influential travel agency.

It was near lunch time, and there was hardly any progress at all. Bijay was getting disappointed till he talked to a fellow next to him trying to figure out if there was any way to expedite his turn. He told Bijay that Bijay's case was different as he had a correspondence from the police station. In the case of that fellow, he was approaching for the first time. When the window was closed for lunch, Bijay requested him to keep his place safe, as he made an excuse to go to the rest room.

In fact, Bijay approached a police officer, who was in a big hurry to go for lunch and was walking fast to go out of the building. Bijay mustered strength to politely show him the letter from the police station. He asked him if he could skip the big line and see somebody regarding the auction notice. Possibly, he could request for further action regarding his application for locating his friend Sajjan.

The police officer just asked him to go back to the line, and take his turn, so that others would not be offended. Bijay felt so sorry and regretted his decision to return to the police station just to waste his vacation time for no good reason. So, he returned to his place and thanked the gentleman ahead of him for keeping his place in queue in spite of a rather long break. When others objected, he explained that Bijay was already standing there with him in line since early in the morning, and had gone out for a few minutes.

At about the closing time, Bijay was fortunate to approach the window, and earnestly hoped that the person ahead of him would take a very little time! It came out to be true, and Bijay was now luckily talking to the same person to whom he had already talked before lunch.

He was impressed that Bijay listened to him and went back quietly in line to take his turn.

Bijay showed him the letter received about 2 years ago from the police station, and the recent auction notice. When the officer asked Bijay's intention, the latter showed his original application with request to locate the whereabouts of his close friend, Sajjan, who used to live in one of the houses in the huge complex. It was now deserted and locked from outside without any security guard for quite some time, and now put on auction by the Income Tax Department.

The officer explained to Bijay that as the tax matter was probably a federal matter, the local police station could not help him at all! He should approach the Federal Tax Office for any information regarding the Auction Notice. The police officer was laughing at Bijay, as he knew that it was several times more difficult to approach them and get any-thing meaningful, unless Bijay himself could bid on the Complex. He knew by looking at Bijay that there was no way that Bijay could bid anything worthwhile! It was about closing time, and he thrashed all the papers back to Bijay, and grudgingly closed the window.

Bijay felt emotionally hurt, but he had expected it. So, he merely took a taxi back to his house, where everybody was expecting him long back. Everybody was glad to see him back. Bijay took his evening tea with snacks and relaxed. His father, however, could see from his face that nothing meaningful came out of his going to the police station. Still he asked Bijay as to how the things went. More important was to know Bijay's next plan, as he was almost towards the last part of his vacation, and he still had a lot to do.

Bijay told his family that he would go for shopping next day with them, and also finish the bank work by getting the latest balances, paying for safe deposit locker etc. This information was urgently required by the U.S. Tax Department. In the evening, he accepted one of his close relatives' long pending invitation. He took the whole family, and his parents were also with him. It was the way things should have been right from the start of his vacation, but he was constantly haunted by his urge to locate his close friend, Sajjan and his family. In the late evening, he packed his stuff, and next morning, he rebooked his reservation along with his family's, as per schedule, as there was no point in his staying back a few more days just in pursuit of taking Sajjan's matter a bit further!

After lunch and a little rest, Bijay was just gleaming through the daily newspaper. His wife had gone to her parents' house along with her children and asked him to join them for dinner in the evening. He wanted to see her family anyway and wanted to say goodbye to them before leaving for USA. His father was taking rest, and his mother was busy in daily household chores. Bijay suddenly realised that the Tax Office was on way to his in-law' house. As it was early afternoon, the Tax Office must be open.

So, Bijay took all the papers with him, took a taxi and reached there. At the Tax Office door, he showed the newspaper ad from their office in the local newspaper to the Security Guard. The latter promptly directed him to go to a particular officer for finding details about the ad. Bijay thought that as it was advertised, there must not be anything confidential about the details on the ad. In his mind, he would be satisfied to know anything about the auction notice. There were a lot of people ahead of him and Bijay wondered as to whether they all were there in connection with the same ad.

Most of the people ahead of him in the long queue were very well-dressed, and had briefcases, probably full of papers, documents etc. Good part was that the waiting time in each case was much less than that in the police station, as the window officer allowed each person standing in line to go inside the office only after initial interrogation. The office inside was very big indeed and had a lot of officers attending to each case exclusively. Later, Bijay came to know that most of them had been asked to appear in the tax office specifically on that day during the business hours regarding tax matters.

Though Bijay, because of a long line ahead of him, was prepared for the disappointment of not being called, to his great surprise, his turn came much faster than he expected. This time, Bijay got an opportunity to see a senior tax officer, and smartly showed him the newspaper cutting regarding the auction notice for the complex. The officer also had the time to listen to Bijay about the unsuccessful attempts to locate his one-time closest friend and classmate.

The tax officer went through Bijay's papers regarding the lengthy form from the police station. He was sympathetic to Bijay for being so patient despite local police authorities not putting enough effort to locate Sajjan and his family. He smartly located the confidential file cabinets regarding the tax matters about the complex and the people staying there. It was not possible for him to go through all the tax files, but he could see the massive taxes due from the family over several years. He could also see clearly the various tax notices issued from the tax office, very lengthy correspondence from lawyers on both sides regarding this long-time dispute.

The tax officer briefly and courteously explained to Bijay that the auction notice in all Indian newspapers throughout the country was the result of non-compliance of taxes due from the complex owners. Realizing it was close to quitting time, and Bijay had already taken a lot of time from the tax officer, he thanked him profoundly and gratefully. The tax officer did not disclose the amount to be paid, nor did Bijay ask him, as there was no way for him to do anything about it, what to say of paying even a fraction of it!.

After all, what mattered to Bijay was not the amount of arrears, but information on whereabouts of Sajjan. He politely asked the tax officer as to how he could contact or locate Sajjan or one of his family members. The tax officer did not see any correspondence from the individual family members, as all letters came from the lawyers representing the family.

Bijay was not sure asking the tax officer for at least one lawyer's name, so that he could possibly give him Sajjan's whereabouts to Bijay. Later, he realized that all this information is strictly confidential, and it was merely a kind gesture and courtesy that the tax officer was telling him all this. It was not that Bijay appeared to him as a potential auction bidder, but Bijay opened his heart regarding the reason for seeing him. The tax officer could see his frustration so far and keen desire to locate the whereabouts of his close friend. Once Bijay disclosed to him that he would return to USA very soon, he was really moved and wanted to help him as much as he could. He was glad that he saw the tax office before giving up. Though Bijay was, nowhere, nearer to solve the puzzle, at least he now knew the reason for his friend's house and the entire palatial complex to be auctioned.

Next day, when he was having tea with his father in the morning, his father curiously asked him about the pending things, and wanted to know as to whether he could be of any help. His father knew that time was moving very fast, and there were so many things to be done before his family and he himself would return to USA. He also knew that he could hardly help him in completing bank matters, shopping, saying goodbye to close relations etc. He himself was very weak and wished his son with family could stay back longer, if not for ever. He knew very well that it was only a wishful thinking which had no reality, not even possibility.

Bijay thanked his father for his kind offer to help him despite being sick and weak. He briefly told him about his plans to finish the pending work, such as getting bank statements, meeting friends and family members, shopping etc. Some important part of shopping was to buy some typical food stuff, like pickles, papads (lentil wafers), long lasting sweets like almond and cashew nut cakes etc., that he could take to USA for distribution to friends. It is indeed unbelievable that these things are so cheaply and abundantly available in India and are rare and very expensive in USA. People who can carry these items with them usually serve these as treats in parties at home and distribute among friends. Usually, there are problems in carrying these items. First, these are subject to Customs' Scrutiny obviously due to health reasons, and second due to baggage limitations.

At that time, his father reminded Bijay about seeing the CPA, who used to prepare and file tax returns for the family, and Bijay particularly. Though Bijay did not have any payroll income in India, he still had some bank interests, stock dividends etc. to report in India. Bijay, on the insistence of his father, went to the CPA, whose address was given by his father first thing in the morning next day. Luckily, he was available immediately, and got Bijay's signature on the tax return form, which he assured Bijay, he would complete soon on getting

Bijay's numbers on interest, dividend etc. Just before leaving his office, Bijay thanked him and anxiously asked him about his help in handling the auction notice in the papers regarding the palatial complex, where his friend Sajjan was residing along with family members.

He simply laughed and told Bijay that there are thousands of auction notices, CPAs and lawyers in Kolkata. It was almost impossible to get any information regarding the auction notice without essential details on the matter. Bijay had the auction notice with him, and he showed it to him. The CPA immediately pointed out the amount involved must be astronomically high. Only the most well-known and very reputed CPAs and lawyers must have been associated with the case. By way of courtesy, he assured Bijay that he would inform his father in case he gets any information while casually talking to his friends, but it was highly unlikely that they would know about it.

Bijay spent rest of his time during his stay for taking care of essential things like shopping, going to banks, helping his parents, particularly in getting them new eyeglasses etc., visiting relatives etc. He requested his father to pass on the required information to the CPA for filing his tax return and mentioned about his assurance to find out about the auction notice on his friend's complex. His father promptly commented that the CPA was too busy to think about it and Bijay should not expect anything free from him.

On returning to USA with his family, Bijay got engrossed in overdue payments, checking important mail, getting ready for his very demanding job etc. Same was the case with his wife, Manju and their two sons, Vikash and Vishal, who prepared themselves to go back to their respective schools. This long-time gap due to their vacation wanted a lot of catching up on the part of everybody and left no time to think about anything else. Life, once again, became extremely busy, as before going to India. For sure, it was clear that it will be several years before they would ever visit India again.

One Sunday morning, Bijay was looking at the papers brought from India. He wanted to keep everything, including bank statements etc. in place. Manju was working on the album. They took several pictures, which would cheer them up and freshen up memories about their parents, friends and relatives on both sides. Vikash and Vishal were allowed to sleep a bit longer, as they were really having very hard time catching up with rest of their classmates.

Suddenly, Bijay opened papers regarding his fruitless venture to locate his close friend, Sajjan. He wanted to file all the papers very carefully with proper notes so that if he ever undertook this meaningless route again, these will be very helpful indeed, particularly his application to the police station requesting their help in locating Sajjan, reply from the police station, newspaper ad regarding auction etc.

By stroke of luck, which was indeed a miracle, he saw a paper, mistakenly given by the tax officer along with other returned materials. Bijay's first reaction was to return it promptly to the tax officer, who kindly gave so much of his very valuable time to Bijay! Before doing it, he wanted to make a copy of this, as it was obviously from one of the lawyers representing the huge complex residents and their businesses. It had the address, phone no. and name of the Company. Bijay wished he had seen it earlier, while at Kolkata, but it was now too late!

Bijay's consciousness did not allow him to keep the paper, he was given by oversight at the tax office. He made a copy with a very heavy heart and returned the original promptly to the tax officer at the Federal Tax Office at Kolkata. Fortunately, he had the tax officer's name, address, phone no. etc. He wrote a covering letter to the tax officer explaining the situation and thanking him profoundly. He frankly admitted in his letter that he made a copy of this without his permission, if it was okay with him. Right now, he could not utilize the letter, but hoped to inform his father in due course, in case he had any suggestion or comment.

Bijay requested the tax officer to inform him if he had any objection, whatsoever, in using the information about one of the lawyers in locating the whereabouts of his close friend Sajjan. The latter very definitely lived in that complex with his immediate and cousin families about 20 years back. Time rolled on, and there was no letter, as expected, from the tax office, nor from the tax officer. Bijay also did not have any time for this rather low priority matter.

He and his family had an extremely busy routine day in and day out. It so happened that he did not have enough time even to write his regular letter (once a month) to his dear parents. Realizing that they would be unnecessarily worried about Bijay and his family for not receiving his regular letter, Bijay decided to contact his parents by a long-distance call. He knew that it would cost him a lot and wished he had written a letter as usual in time. Any way, it was too late to think about all these!

On a fine Sunday morning, when Bijay was relatively relaxed, he called his parents. They indeed were very, very concerned, but were very happy now to know that everything was well on both parents' and Bijay's sides. Bijay took the opportunity to briefly mention about the lawyer's letter, mistakenly given by the tax officer. His father promptly advised Bijay to send a copy of it to him in the next letter. Though he did not know at the spur of that moment as to what he would do, he assured Bijay he would give a serious thought, as it would be the first solid evidence with him for the auction notice on the complex to pursue the matter further.

Bijay took his father's advice very seriously and promptly sent a copy of the letter to him along with his regular letter. When his father got it, he was stunned to see that it was from one of the topmost law offices in Kolkata. This law firm was known as handling only multi-million rupees (Indian Currency) cases. As much as he knew from his friends, it was in the posh Dalhousie Square, and was staffed only by very reputed British Lawyers from UK. Being a man of humble means, he did not even know as to how to proceed or talk to somebody about the matter, which was still bothering his son, Bijay, though he had no contact with his onetime friend, Sajjan, since long!

Things gradually settled down with no action on Bijay's or his father's part. Suddenly, his father had an idea to contact Bijay's CPA, whom Bijay had seen just before leaving for USA. First, he called him about the lawyer's letter to Federal Tax Office in Kolkata refuting their charges for the unpaid taxes, penalty, interest etc. The CPA told him frankly that he would need the actual letter to do anything, if at all feasible! Bijay's father made it clear right then that the CPA should not put much time, as Bijay or him himself, could not pay anything regarding this matter. The CPA did not have anything to say till he got the letter, at least a copy. So, Bijay's father, next day, took the letter himself to CPA's office, requested him to make a copy, and hand over the letter back to him for further reference!

Time rolled on with no news from the CPA, who was indeed very busy with his normal professional work. He used to prepare tax returns for middle or low income individuals or families or businesses, just like Bijay's, and helped them in tax planning for the coming year. His charges were very reasonable, the reason Bijay and his family chose to go to him for years after years. As he was associated with Bijay's father for a long time, he, at his own, one day, started looking at the copy of the letter from the lawyer to the Federal Tax Office in Kolkata, strongly refuting all the charges etc. on behalf of Sajjan's business empire.

It indeed was a very famous law office not only in Kolkata but throughout India. CPA did not take much time to understand the matter. As the letter was dated about 8 years ago, it was obvious that the dispute was probably going on for a very long time between Sajjan's and his family's business empire on one side and the Federal Income Tax Office authorities on the other. It was certainly a case of huge tax evasion by the business empire, and the responsible party did not pay despite repeated reminders etc. The Federal Tax Authorities forced the Complex residents to evacuate it and the entire business was closed, possibly moved to another place as a different enterprise.

The CPA decided not to call the lawyer's office, as he had no connection with any entity. So he called Bijay's father and informed him about his inability to help Bijay and his father on this matter. If he puts enormous time and effort by going to the lawyer's office or contacting him, he will merely be turned down for lack of authority to investigate on the matter. All that he could ask the lawyer was possibly to find out about the final option for the auction notice appearing in the news-papers throughout the country.

Obviously, the tax arrears were huge, and the business refused to pay. So, there was no other way than to evict the residents with proper notice. The families must have moved to some place, possibly unknown to the lawyer. On the other hand, the lawyer or lawyers representing Sajjan and his family must be in touch with them, particularly as auction notice is a significant information for them also.

In any case, as CPA indicated to Bijay and his father, it was a time and cost-intensive project for him to pursue any further. Bijay and his father did not want to spend any money for the unnecessary headache for them except for Bijay's mental satisfaction. So, Bijay decided to close this chapter at least for the time being, till he would visit Kolkata again, possibly in a couple of years. It was obviously the best decision for Bijay, as he could concentrate on his job. His wife, Manju, was also happy to see her husband back to normal life. Their two sons got back to all activities outside of normal studies.

Time went fast and Bijay was ready with his wife and children with plans to visit the aging parents on both sides. Obviously, it was very expensive, and almost unaffordable due to rising costs of airfares etc. On the other hand, they were constantly under stress, thinking about, their parents, who were often sick and wished to see the whole family.

Day came when Bijay flew with his whole family from Dallas, USA to Kolkata, India. On reaching Kolkata, he again passed through Sajjan's complex, but he did not wish to stop in the middle of night, and his family was with him—all extremely tired and exhausted due to jet leg, long flight etc. Passing through the Complex, however, triggered his memory about his close friend, Sajjan, and his futile efforts so far over the years to locate him. He was reminded about his CPA's efforts, which might have been useful, had he pursued the matter, but he did not wish to spend money. He reconciled himself in his mind, and on reaching home, he took enough rest to overcome jet leg due to huge time difference of about 12 hours.

While taking tea with his parents, wife and children, Bijay started looking at the copy of the lawyer's letter to Federal Tax Office, though he was not fully convinced about possible advantage of going or contacting the lawyers' office. First, the lawyer would be shocked to see the copy of the strictly confidential letter in Bijay's hands. Second, what is Bijay's intention by keeping it? Third, why was a stranger like Bijay, unknown to him, got concerned about the auction notice? Fourth, was Bijay going to bid, and if not, he better, take his hands off the matter. It might put Bijay in trouble, who could not afford to have any blemish on him in any case, be it Sajjan, his close friend or/and his own family member.

After a lot of thinking, Bijay decided not to call the lawyer's office in the prestigious Dalhousie area of Kolkata. It was indeed a big legal issue, and even if somebody picked up Bijay's phone, immediate advice would probably be for Bijay to go to the lawyer's office. So, he decided to go there one day instead of calling by phone. Time was running fast with so many errands to be taken care of, before leaving back to USA.

Yet, Bijay took time off to see the lawyer, and enquire about his closest friend Sajjan. At heart, he was very hopeful, but he had several doubts as to whether he would be able to achieve anything. On reaching the

Office Building, the security guard enquired about Bijay's purpose of going there. When Bijay explained, the guard sent him to one of the receptionists, who asked for Bijay's identity etc. Bijay smartly showed his U.S. Driver's License, as he had no other identity with him at that time. He did not carry the U.S. Passport with him. He just wanted to speak to somebody in the lawyer's office.

To make his case forceful, he showed the copy of the letter from the lawyer's office to Federal Tax authorities. The receptionist called the office and on getting their ok, Bijay was allowed to go to the ninth floor, where the lawyer's office was. As soon as he got off the elevator, he was met by an Office Secretary. She escorted Bijay to the Office and asked him the purpose of his visit. Bijay showed the copy of the letter and requested her to put in touch with somebody, who could tell him about whereabouts of his friend, Sajjan.

As the letter was dated long ago, the Office Secretary told Bijay that as per policy, records on the closed or settled cases are reverted to their London Office, and it was very time-consuming and cost intensive to get the papers back to Kolkata Office.

Seeing that Bijay was almost at a breaking point, he showed her the auction notice on the Complex, which appeared about 3 years ago in all newspapers. The lady promptly asked Bijay as to whether he was the Complex Bidder. When Bijay denied, the lady, point blank, asked Bijay about his role in the whole matter, particularly in possessing a copy of a very confidential letter from that Office. Bijay had to admit that his role was just to locate his close friend, and in that process, he got the letter from a tax officer in the Federal Tax Office by oversight.

The lady Secretary started doubting Bijay's intentions and asked him to wait. A high-level Security Officer called Bijay soon and asked him to surrender the letter to Federal Tax authorities. If he did not comply, he would be arrested on charges of theft of a confidential document from the Tax Office.

Bijay had no alternative but to surrender the copy of the letter to the Security Officer and sign a letter stating that he had no connection with the Complex case and was just trying to locate his missing friend, whose name did not appear anywhere in that Law Office.

The Federal Tax Officer, whom he saw long back, told him that he could not see Sajjan's name anywhere. Thoroughly disappointed, Bijay had to return home losing practically the whole day and achieving nothing. In a way, he considered himself lucky that he came out of the Law Office without his name appearing anywhere, otherwise it could have been a serious matter against him and could have cost him his current U.S. job.

On returning home, when his parents asked him about the day, he somehow avoided mentioning his unfortunate episode on the matter of locating his old friend. His wife and children were waiting for him the whole day, and he decided to take them to a movie for change of mood. Bijay, particularly as a result of his experience in the Law Office, firmly decided not to pursue Sajjan's matter further. Now, he did not have a copy of the letter from the lawyer's Office. His only document was the Auction Notice, which was obsolete now. He thought somebody must have bid on auction and might have got the complex by paying the bid amount.

Time went fast, and he was ready to leave Kolkata with his family. Just at that time, his father gave the copy of the lawyer's letter that Bijay had sent to him for CPA. Bijay's father preferred to return that to Bijay, as he had cleared all the papers at home. So, Bijay unwillingly accepted the copy of the letter that he had sent but saw no use of it. In view of his very bitter experience, he did not want to think about it anymore. He did not tell his father about his very bitter experience in the famous lawyer's office.

On returning to Dallas in USA, everybody once again reverted to routine- Bijay and his wife with work and children back to schools loaded with homework and sports activities like soccer etc. On one long weekend, Bijay decided to take the whole family to Austin, about 200 miles from Dallas. This, being nice weather, everybody liked the idea. As they went by car, they could move around Austin.

Bijay with his family Texas State Capitol Building, President Johnson Library, University of Texas campus etc. They also had the opportunity to go to the famous Barsana Dham – a Hindu Temple with a lovely surrounding around the temple. In the temple, suddenly Bijay came across a familiar face- a lady, Vinita, who was the wife of one of his long-time friends, Vinod. They greeted each other and introduced their families. Bijay asked the lady about her husband. She told him that he was visiting India on a spiritual mission. Spiritual mission and Vinod? It looked like a joke to Bijay, as Vinod never had any interest other than work, and family.

Later, Vinita invited Bijay and his family for dinner to see her other family members, two sons, in their house, which was in a very nice neighborhood of Austin. Next day, Bijay happily went with his family to their place, and met all the family members. Bijay was happy to see them all. During conversation, Bijay once again talked about his friend, Vinod, who was possibly visiting India for a short time. He was curious as to his going alone, leaving his wife and family in Austin. How did he manage to take time off work?

He was told that India has become almost a permanent place for his friend, as he was engrossed in spiritual life only. His friend, several years back, came across a monk, who influenced him a lot and apparently changed his life completely. It was indeed very interesting, and yet very hard for Bijay to believe that one would give up his family responsibilities altogether. It must have hurt the lady, his devoted wife, very seriously. As they all sat down for dinner at the dining table, he saw a picture of his friend. He was bowing to a monk with folded hands and head down, and Bijay could not see the monk clearly from a distance.

On finishing dinner, though it was getting late, Bijay quietly walked to the wall, where his Austin friend's picture was hanging. On seeing it closely, the monk strangely appeared to be a familiar face to Bijay. He tried to remember, as hard as he could, but could not put together his stray thoughts about that familiar face. He went to the hotel with his family and took rest in the night. In the morning, he started thinking about his Austin friend, his family and the monk to whom his friend was bowing with great devotion.

Somehow, the monk's face appeared familiar to Bijay, just by coincidence, but he still could not put all the strange thoughts in his mind together. Just for curiosity, he called his friend's house, and talked to the lady thanking her for inviting them for a very nice dinner. He also enquired about his friend's possible plan for the next trip from India to Austin. Nothing was certain. So, Bijay requested his friend's wife to inform him about his friend's next visit to Austin, whenever it was.

Bijay took his friend's phone number in India just in case he decided to call him. There was no address for him, as he travelled with the monk, whose itinerary changed every day. All that was perplexing Bijay's mind was somewhat familiar face of the monk, and his influence on his Austin friend. Can he learn more about both? Later, he asked himself as to how it concerns him after all. He had enough problems of his own, though at the back of his mind, the monk's face was challenging his memory for no reason whatsoever.

Time went fast- months after months. Suddenly and surprisingly, Bijay got an unexpected but welcome call from his friend Vinod in Austin. He was visiting his family there for a very short while. His wife, Vinita, had told him long back about Bijay's request to call him whenever he was in Austin. She had given him his Dallas address and phone number, but very understandably, Vinod could not call or contact him earlier. Bijay did not have any way to call or contact him in India either. He was always on the move, and even his own family did not know his whereabouts.

Bijay came straight to the point of his Guru-the monk in the picture that he saw in his house with Vinod bowing to him with folded hands. Bijay was anxiously listening to his friend, Vinod. According to him, this was his biggest achievement of his life to have found a Guru (monk) like him.

Even without asking, Vinod frankly told Bijay that the monk has exceptional super spiritual powers, and his philosophy, teachings, way of life etc. are all exemplary. He admitted that he never had so much peace of mind till he came across him and became his staunch devotee. Bijay could not believe such a big change in his friend, Vinod, who was just like any other professional. As much as Bijay knew about Vinod, he was working very hard maintaining his family, and taking care of his obligations without burden on his wife or other family members. He had an excellent reputation in his field and was a recognized expert.

Now, that Vinod has changed so much, how is he managing family expenses at this critical junction of his life? It was indeed a big puzzle for Bijay, but he wished to stay away from personal matters as far as possible. When Bijay asked Vinod about his future plans, he firmly said he was definitely returning to the monk, and was making only a short visit to Austin to see his wife, Vinita and two sons. About family expenses, he mentioned about Vinita still having a good job, his savings, etc. The house was already paid for. Admittedly, he felt extremely sorry that he could no longer help and support the family, who according to him, have realized that they are at their own. Everybody in the family missed him a lot, but he was adamant on his decision.

Bijay, in his own mind, thought that it was a classic case of the clever monk hypnotizing simple Vinod. Otherwise, how could such a big change come at this age, when Vinod was still young, and had a lot of years to live. Before disconnecting the phone, Vinod asked Bijay as to whether he would be able to accompany him and see the holy monk himself. Bijay had to admit that though he was impressed by his friend's observations about the monk, he personally would not be able to accompany him. He took his address- Anand Ashram, Pune,

India. He also took the monk's name- Anand, and obviously the Anand Ashram (shelter, where the monk lived) was named after the monk Anand.

Though Bijay turned down Vinod's offer to accompany him to India, he somehow felt very uneasy thinking about some sudden changes in him. These changes were adding to his complexities, like aging parents on both his and his wife's sides, locating his close friend Sajjan, demanding job deadlines, family pressure, after-office night courses etc. It was too much for Bijay to see Vinod's sudden change from a normal, apparently a very successful professional life, to a hard spiritual life. He was leaving his family, particularly his lovely and devoted wife, Vinita, who was still young. Was he migrating to India for good just to stay close to his spiritual teacher, living an austere life?

In Austin, Vinod had a normal life with all the materialistic comforts, but he had chosen a hard life himself. It was indeed difficult for Bijay to understand and comprehend. He was really concerned about Vinod's family pulling on without him- a big support indeed for the whole family. Another intriguing factor, constantly bothering Bijay, was the monk's picture in Vinod's house. Though he was recollecting and challenging his mind and memory, he was still unable to put the things together. He strongly believed that the monk's face was familiar to him, and he had seen the monk before his monkhood, but he could never be certain about his guess work of knowing him or seeing him ever.

It was not easy for Bijay to see the holy monk clearly, as his body, including his forehead, head etc. were all covered by a white chadar or white sheet. Only his eyes, and a little bit of face were visible. Rest was all covered. In the morning, Bijay had to rush to work, and complete a lot of work, as the deadline was very near. So, he took his mind off from the complexities, but when he returned home in the night after evening/night courses at the University of Texas at Dallas, he was again thinking about the whole matter, particularly about the monk. He told his wife about all these, and she suggested to him to relax and not worry about unnecessary things.

Very honestly, in the back of his mind, Bijay had a complex maze of three problems to be solved. First, locating his friend Sajjan, second finding about the sudden and completely unexpected change in his Austin friend, Vinod, and the third about finding more about the monk, who not only mesmerized Vinod, but the feeling that he created in Bijay's mind, as if his face was already familiar to Bijay. Most of the time, Bijay was engrossed in heavy workload, evening classes etc. etc. In the weekends, sometimes, he could not help thinking about these problems again and again. The way Bijay has known Vinod is that he was a very practical person, who avoided unnecessary discussions on topics like politics, religion etc. He was always busy with work and family affairs.

In about a year or so, when Vinod called him from Austin, Bijay invited him to visit him in Dallas and stay with him. Vinod and his wife accepted Bijay's offer and drove down to Dallas and stayed with Bijay and his family. As there was plenty of time to talk about Vinod's new life, Bijay point blank asked him about the sudden change to stay near his spiritual teacher, Anand in India. Vinod's wife, Vinita, insisted him a lot not to leave home, as it was not, at all, easy for her to manage the family in his absence.

First, she thought that he would visit Anand in India occasionally, and would turn back to his normal life, but now, despite repeated insistence from his family and close friends, Vinod had firmly decided to stay there permanently. This was Vinod's last trip to Austin, USA. As the rest of the family had to be taken care of, Vinod's wife, Vinita, could not accompany Vinod. Luckily, she was nicely employed, and her two sons were excellent college students.

V.

NEW LIFE

Vinod felt grossly immersed and happy in the spiritual world because of his Guru Anand. Vinod came in close contact with him when he was visiting Pune, India, and was very impressed by his teachings. Bijay wanted to know more about Anand, but Vinod did not know anything about his life before monkhood. Neither did the monk ever speak about it, nor did anybody dare and care to ask him. His sermons apparently touched hearts of anybody listening to him. Being lunch time, they all discussed about as to how the whole matter evolved and changed Vinod's life.

It is interesting that Vinita's parents lived in Pune, India, and when Vinod, Vinita and their sons visited them, out of Gratitude to the Almighty, they wanted to take the whole family to Anand Ashram, which was also a very scenic and peaceful retreat, apart from being a big spiritual heaven! The whole family planned to have a big picnic there. They all really had an excellent time over there, and were ready to return. Right at that time, they saw a big group of people patiently waiting to see somebody. As they could not take the car out till the big crowd was cleared, they decided to enjoy fresh air in the open. In a few minutes, they saw a white-clothed, lean and thin monk gradually walking towards the crowd. Vinita's parents also wished to offer their respects, along with all in the family even though they knew nothing about the monk.

First reaction from Vinod was negative, as he did not like to hear or participate in any spiritual matter. Spirituality and Vinod were diagonally opposite. When Vinita and family insisted hard or forced him, to go along with them, he also very unwillingly followed them. He saw that the holy man's face was very cheerful, his aura was indeed very bright, and he was smilingly giving his blessings to all of them. They were all paying respects with folded hands to the monk one by one, as everybody was standing quietly in queue,

When Vinod's turn came, he also bowed to him with folded hands, and the monk, as usual, gave him his blessings flowing like bright sun's rays. Vinod was wonder struck by his bright aura, cheerfulness, love and kindness all around. Vinod had so far tried to avoid such people, and this was the first time he came across a person face to face who impressed him a lot right at the first sight. He asked Vinita's parents as to whether he could see him again. Her parents and she herself were really surprised, and assured him that they could visit him sometime in near future before Vinod's departure to USA. Everybody in the family that day was pleasantly surprised about Vinod's request to return.

Time went really very fast, and there were tons of formalities like visiting close relatives, shopping etc., and nobody remembered about Vinod's request to see the monk again. One day, when everybody decided to go for shopping, Vinod took family's permission to visit the beautiful Anand Ashram, and possibly the holy monk. He was not sure that he would be allowed to see him. He distinctly remembered the big crowd that day when everybody was anxiously waiting for a personal glimpse and blessings of the monk.

Vinod took a cab and proceeded that way, knowing fully well that the monk may be too busy to see anybody individually. The cab driver took him straight to the famous Anand Ashram. Vinod did not know about the exact place where he could see the holy monk. He started walking and kept walking without much success. So, he finally asked a passer-by about his dilemma. The gentleman told him that the holy monk was giving sermon in an open space.

The gentleman also told him that he could have the holy monk's glimpse immediately after the sermon. He cautioned him that he may not be able to spare too much time for Vinod, first as he did not make an appointment with the Ashram Authorities, and secondly as the holy monk would proceed for lunch and rest before he sees anybody in the audience. Vinod did not want to give up, as if his inner strength kept him hopeful about seeing him. He walked to the hall, where the holy monk was delivering his lecture, which was probably coming to a close in a few minutes.

The holy monk was emphasizing on everybody to pay heed to the fact that death was inevitable for all human beings, and nobody, including near and dear ones, could offer any protection from death. So, it was important for everybody to realize the truth, and give-up any attach-ment. Body must go, as it is perishable. Soul is immortal and it will re-main forever. These are two completely different entities. Everybody should strive to look into the inner power of soul, and make a strong resolve to achieve everlasting bliss and happiness. These words were piercing Vinod's ears, heart and mind, and he was trying to understand the meaning, which was too deep for him to comprehend. He, just like most, was circling around materials, bodily comforts, money and never ever bothered about anything like soul.

Immediately, after the sermon was over, Vinod just intruded into the crowd to see and get the glimpse of the holy monk. When he was following the monk, he requested him to spare some personal time for him, if possible. The holy monk, even though he had a very tight schedule, agreed to see him when he was proceeding for lunch. Vinod, probably, came across the kindest person on earth, and had no word to thank him for his generosity.

Though very lean and thin, the monk was walking astronomically fast, and it was hard for Vinod to catch up with him. While walking with him, Vinod was trying to impress upon him that he was from USA, but the holy monk was merely advising him to take a fresh look at the spiritual world. He told him that giving up materialistic world, particularly family with near and dear ones, was not easy, but according to him, the main source of sorrow in this world are attachment and violence. Violence is committed not only by body, but also by words and thoughts.

There is no end for sorrows except by introspection and taking the path of non-attachment and non-violence. Each of us is lagging behind, and it will undoubtedly be too late soon! The only solution is self-realization. Each word from the monk's mouth was so powerful and so inspiring that Vinod lost himself. He did not realize that the monk was near his residence, where he had only a little time to complete his lunch, take some rest and prepare for delivering his teachings in depth to his followers and fellow monks.

Vinod never believed in spiritual discourses arranged in Austin from time to time. First, he was so busy and then he had too many other problems to be taken care of. There was a constant influx of problems one after the other. When Vinita and his two sons requested him to spare some time on very special occasions, he forced himself and ignored celebrations. But the situation was completely different now. It was a miracle indeed that he came to Anand Ashram at his own initiative. When he came in contact with the holy monk and was walking with him, each word was flowing like nectar from his mouth. Each word came so smoothly and kindly that it pierced Vinod's ears and touched his heart. He was never influenced so much in such a little time by anybody earlier. As the monk moved away from him to take care of other things, Vinod just offered his best and humblest respects to him, bowing his head with folded hands.

It was a sudden change in Vinod's attitude, and it was so spontaneous and sudden that he himself could not express it. He could just feel it. He missed so much and felt extremely sorry! There have been various instances, where people have miraculously changed either by remembering their past or encountering some inspiring and godly persons or by listening to some noble people like the holy monk. In Vinod's case, it appeared to be the third possibility.

Vinod immediately realized that he was in Bijay's house in Dallas, and had finished lunch, just talking to all by himself alone. He could not help, as he was completely lost in his first encounter with the holy monk Anand in Anand Ashram. He could not even tell Bijay that he became completely restless during his last few days at Pune before returning to Austin, USA. Even on returning, he could not concentrate on his job, family responsibilities, his commitments etc., which he enjoyed and was doing his part for the last 25 years, before going to India.

Vinod's family could distinctly see the big change in his attitude, but sincerely wished and hoped that he would eventually forget about the monk completely. Vinita kept on emphasizing on him that the monk's teachings were not real world, and Vinod could not afford to prolong his 'madness' any longer. Even the children reminded him about his promises to them regarding their summer camp, vacation plans, etc. They all realized that their world would turn upside down without Vinod.

Nothing worked on Vinod despite everybody's best efforts. How could he ignore his family of four? His boss at job kept calling Vinita, and kept on warning her that he would eventually lose his job, as he was supposed to complete his regular assignments ! Even his close friends could not convince him. Vinita took him one day to Barsana Dham - the most peaceful and spiritual retreat in Austin and Vinod had the opportunity of seeing and talking to the pastor over there, but that also did not satisfy him! He was completely engrossed in Anand monk and his teachings. Every word from the holy man was engraved in Vinod's heart and mind.

Later on, during the day, when everybody was having tea and snacks in Bijay's house, Vinod again asked Bijay to follow him to India. Bijay, would have liked to see the holy monk himself. His glittering face in the picture in Vinod's house appeared familiar to him. He just could not take time off as he had too many activities going on, such as evening computer science classes at the University of Texas at Dallas, deadlines on several work-related projects, etc. etc.

Bijay thanked Vinod, wished him all the luck and requested him to keep in close contact with him! Even Bijay and his family would miss him a lot, what to say of his own family members, but Bijay realized that Vinod was at a point of no return whatsoever! While taking tea, Vinod told that he had resigned from a well-paid job in AT&T, where he worked for almost 25 long years! He told all about his savings, retirement settlements etc., and entrusted all accounts to Vinita. It was very painful for Vinita, as she still believed he would change his mind at least for the sake of his bright sons. It was all a wishful thinking and Vinod did not regret his decision to leave them forever!

Bijay, realizing that this could very well be Vinod's last trip to Dallas, he assured his best to Vinita and children, requested them to keep in touch with him, and not hesitate at all to inform him in case of any problem. Thus Vinod's stay with Bijay was almost coming to a close- it was indeed incomprehensible and the most perplex problem for Bijay and his family to understand such a devastating change from a devoted family person to a lonely life in the pursuit of something uncertain and unknown!

In the evening, when Vinod left for Austin with his family, Bijay, reflected on the whole matter, and for now, at least try to understand the reason for his leaving his family unexpectedly in search of lasting peace, as Vinod called it! What was lasting peace in this world.? At least for Bijay, it was a big puzzle! Was Vinod leaving family for the holy monk, Anand, the only solution? Vinod was trying to find lasting peace for himself without caring for the rest of the world around him! Was it not being selfish, if not stupid?

Bijay's own mind was still perplexed with two other problems- finding whereabouts about his close friend, Sajjan, and the apparently familiar face of the holy monk, whose picture was hanging in Vinod's house. It was admittedly not at all a clear picture, as the monk's face was covered with his white Chadar.

Bijay and his family rolled back to their very busy life as before. Their life was so busy that there was no room to think about anything else other than the critical matters. Only in the weekends, there was a little time to talk to friends, family etc. Bijay made it a point to call Vinita, and came to know from her that they are gradually learning to live, rather pull on, without Vinod. What an irony of fate? One small incident changed the family life forever! Had Vinita known about it, she would have never visited her parents, Anand Ashram etc., but it was too late now! She felt guilty for all this mess and regretted every minute. Her parents were blaming themselves. They felt everything was lost forever because of their stupidity!

Vinod used to call Vinita and his sons occasionally, not more than once a month, and apparently, he was doing fine! His constant association with the holy monk Anand not only helped understanding his teachings more deeply, but people started calling Vinod a Yogiji (a devoted person- devoted all the time to the holy monk without caring about anything else!). Gradually, Vinod started forgetting about his family, as he was completely engrossed in serving the holy monk as the most obedient and devoted disciple ever!

In about 3 years, Bijay had accumulated enough vacation, and was asked by his employers either to use or lose his hard-earned vacation time! Bijay and his family decided to visit their old parents in Kolkata, India. This, appeared to be the best utilization of their vacation time, and parents on both sides welcomed the idea! Again, Bijay passed through his school friend's house on his way from Dumdum Airport to his parents' residence. As it was midnight, everything was dark near the Massive Complex!

Bijay was sure that somebody must have successfully bid and occupied the palatial Complex! The new occupants may not even know anything about Sajjan and his family, his cousins and close relatives! It was hurting Bijay to even think about his fruitless efforts to locate Sajjan, in spite of his best efforts during his previous trips. The last episode, in the lawyer's office, involving security etc., almost got him arrested due to possession of a confidential letter, interpreted as having been stolen by Bijay! Though the real truth was far from it, Bijay did not want to be trapped in legal matters!

Over the weekend, Bijay suggested to his wife to undertake a one-week trip and stay near Bombay Airport, from where they all were flying back to USA. Bijay hesitated to tell his family frankly that he wanted to see his Austin Friend, Vinod, and his preceptor, the holy monk, Anand, in Anand Ashram, if possible. He was trying to recollect his glowing but faded face in the picture hanging in Vinod's house. Anand Ashram, luckily, was near the Airport. The family liked the idea, and they all flew from Kolkata to Bombay just a week before their departure to USA. They were excited about their shopping etc. in Bombay, which apparently was and even now is shoppers' paradise.

Bijay and his family stayed in a hotel near Bombay Airport, and spent most of the first 4 days in sightseeing, shopping etc. His wife and sons wished they could take some rest for a day, and take it easy, as the weather was very nice, and hotel facilities were superb. Apart from being neat and clean, it provided excellent food, and most other facilities, like swimming, exercise, recreation etc.

Bijay liked the idea and told the family about going to Pune by car. Pune is not too far from Bombay Airport, and they all decided to return late the same day. On reaching Pune, everybody wanted to take some rest, where they could relax and get some fresh, crisp and cool air! Bijay suggested the beautiful Anand Ashram, which was not too far. He remembered that Vinita's parents lived in Pune, but he did not know their address, nor did they inform them about their coming to Pune in advance. They proceeded by the chauffer-driven rental car straight to Anand Ashram. They all refreshed themselves, had hearty lunch, and took some rest. It was almost late afternoon. Bijay told them about seeing Vinod, his Austin friend, who was supposedly living there. Everybody remembered him, and very much liked the idea, particularly having come that far!

On enquiring from the Ashram Office, they not only confirmed Vinod living there, but also got direction to see him. Bijay, without delay, instructed the driver, to take them straight to the place of Vinod's residence. On reaching there, they all saw Vinod in deep meditation. It was completely a surprise visit from them, and Vinod had no idea of their coming. He was not at all prepared to see them so unexpectedly. At about 4 PM, some of his devotees visited him, and greeted him as respectable Yogiji Maharaj.

Nobody except the Ashram Office knew about his real name, Vinod. Vinod had completely changed with beard, garland around his neck, wearing simple Kurta (white shirt) and Dhoti (loin cloth). His residence had practically nothing except books, prayer materials, and plain unpadded carpet with a meticulously washed white bedsheet (chadar) without any pillow etc. He had deserted all material comforts long ago, and devoted himself entirely in the service of His Guru, the holy monk Anand. Vinod's face was glittering with a shining white aura around his face! He looked healthy and quite cheerful.

After Vinod's or Yogi Maharaj's regular devotees left him, Bijay and his family, slowly approached him, and respectfully greeted him with folded hands. He was completely surprised to see them, but he was very happy to see them. He immediately asked them about their plans. Bijay and his family indicated that they had to return to Bombay in a couple of hours. At that time, a devotee came and respectfully whispered to Yogiji Maharaj about seeing the head monk, Anand, immediately! Yogiji asked Bijay and his family to accompany him, so that they all could get at least a glimpse or darshan of the holy monk, who was extremely busy and constantly surrounded by at least a hundred followers!

It was indeed very kind and thoughtful of Yogiji Maharaj, to take them along with him without any appointment with the holy monk! He briefly introduced Bijay to the monk as his close friend in USA. He never mentioned his name, as it would hardly mean anything to him. Bijay's family respectfully bowed to the holy monk, Anand, with folded hands.

Seeing them with Yogiji, the holy monk Anand raised his forehead, and while he was giving blessings, he spontaneously called Bijay by his name! Though he had turned thin and weak, his memory was so sharp that his mere look rekindled all the forgotten years in Bijay's memory. Bijay still was not sure. He was wondering as to whether it was a dream or reality! Did he hear correctly? Could Anand monk be his long-time close friend, Sajjan? As the monk had a very busy schedule, he merely blessed Bijay and his family, and proceeded for an intensive session on very deep discussions with his followers, devotees etc., including Yogiji Maharaj.

There was no time to say good bye to anybody, and Bijay still was not sure about his guesswork on Sajjan, who was no other than Anand Muni (holy monk). Bijay knew his restlessness, and told his family about his dilemma about Sajjan vis-à-vis Anand Muni. What was surprising was that though he was called by name Bijay by Anand Muni, he never showed any excitement about meeting Bijay after such a long time! Nor did he ask him about his plans, what to say about his request to Bijay and his family to stay there for some more time.

How could the holy monk Anand Muni know Bijay's name? Very definitely, when Yogiji Maharaj (Vinod) introduced Bijay and his family in front of some followers and visitors, he never mentioned Bijay's name. Could it be that Yogiji Maharaj talked to him earlier casually, but Bijay never ever contacted Yogiji (his old friend, Vinod) in India. Even if he mentioned about him just casually, how could the noble monk guess correctly that he was Bijay. He just called him by name, but never showed any excitement or pleasure, as he had no time at all!

Thinking about a distinct possibility that the holy monk was really his one-time, closest friend, Sajjan! But how was he so different? His face was glowing with a bright shining aura, and his body was very,

very slim. Did he lose a lot of weight because of intense penance by observing long fasts, may be for several days? How could a bulky man, Sajjan, who was always overweight, turn to such a state? Belonging to the famous 'Mica King' family of India, and happily married, how could he become an ascetic? Bijay was so perplexed, as if he was now more restless, and engrossed in uncertainties than before! He was waiting to hear from him that he was not what Bijay was thinking!

Day was nearly over, and it became dark. Disappointed and thoroughly confused, Bijay and his family, with only 2 days left before their departure from Bombay to Dallas, did not have time to say good bye to Yogiji Maharaj, his friend Vinod in Austin. What would Bijay say when he would face Vinita and family in USA? Bijay was not at all sure about his coming back to India in the near future!

Somehow, Bijay and his family decided to return to their hotel, where they already had reservation for a week, out of which only 2 days were left with a score of urgent matters, to be taken care of, before their departure. Though Bijay returned back to the hotel, his mind was circling and constantly absorbed thinking about Anand Muni, and knowing a little more about him before departure to USA. He was not sure about his next step, if any!

Next day, Bijay and family had already planned to leave for Juhu Beach, immediately after breakfast, and return to the hotel late in the evening, after finishing a visit to the Goregaon Milk Diary facility. As soon as Bijay reached Juhu, he became so restless that he told his family that he wanted them to enjoy both Juhu Beach and the famous Diary facility, and then return to the hotel. He wished he was with them all the time, but he took the courage to tell them frankly about his plans to see his friend Yogiji Maharaj, and if possible, Anand Muni with him to know a little more about his pre-monkhood life! He was not sure about success in his plans, but he wished to give a sincere try! He remembered Vinod telling that nobody knows about Anand Muni's life before monkhood, as he never talked about it, nor did he encourage anybody to think about it. How will such an adamant and strict personality reveal it to Bijay, even if he saw him face to face?

When Bijay reached the Great Hall, which was heavily crowded and packed by the Monk's followers, devotees, visitors etc., he had to stand way behind, from where he could hardly see Anand Muni. He could, however, distinctly hear his sweet and kind voice, which was very clear and loud. He was preaching that this life is the result of past and present Karmas (deeds). Whatever one does, his future is shaped by his actions, and actions alone! One should always think that death is inevitable. Nobody, even wife, son. daughter etc. can protect him from death. He is always alone, and every day is bringing him closer to death.

The only reason for sorrow is attachment for money, body, wife, son, daughter etc. So, one must try to overcome the feelings of attachments with near and dear ones. Bijay was so lost in the holy man's sermon, particularly about own family, and entirely different emotions of care, love and affection etc. He was so engrossed that he forgot about time. He had come here for a purpose, and he must fulfil it, if possible. He could not locate Yogiji anywhere in the Hall, which was completely packed with people!

Time was going so fast, and Bijay could not even ask anybody in the Great Hall about the program that day, as there was pin-drop silence despite more than 1,000 attendees in the Hall! He wished Anand

Muni's sermon is over by now, but it was not. He wished that he could somehow locate Yogiji Maharaj at least. Why was there complete silence, and there was no movement at all! If he could locate Yogiji somehow, possibly he could have some luck to see the holy monk face to face before his departure for USA.

That day was the only day, when he could come there, as next day, he must do some shopping for himself and rest of the family. He did not know about returning to India, if ever! The trip to India is very, very expensive indeed and almost unaffordable! Besides shopping, he must confirm reservations on return journey for himself and family back to Dallas, return the rental car, hotel payments etc. They had early morning flight, and he must settle all these errands before calling off the day, even if it is midnight!

Bijay was almost getting disappointed, and impatient. He did not see any help from anybody, particularly as it was very quiet, and he could not dare to whisper to anybody! Even if he would go to the Ashram office, which was rather far, he may not have much luck. He decided to continue to listen to the monk, who was almost coming to his closing remarks on his sermon that day. He was emphasizing that sorrow was due to violence and attachments, and is created not by others, but by oneself. It was very difficult for Bijay to comprehend! Bijay knew from his last visit's experience in Anand Ashram, that the holy monk would immediately proceed to his room, have quick lunch and brief rest before embarking on very intensive studies with his learned and devoted disciples. The whole program was pre-planned, and would continue till late in the evening. Last time, it was so late that Bijay and the whole family gave up, and had to return to the hotel for dinner, rest etc.

As for that day, when the sermon was over, the attendees stood in a big queue to have the holy man's blessings through a momentary glimpse (Darshan), and coming face to face with him. It looked like impossible to have his personal attention, as the queue was really very long, though very disciplined and quiet. Bijay also stood in the queue, and while he was waiting for his turn, he looked carefully for Yogiji in all directions of the Hall, but could not locate him. Did Yogiji decide not to attend the holy sermon, and do something else as per Anand Muni's instructions? Everybody was passing by the monk very quietly.

When Bijay's turn came, he also bowed to him, knowing fully well that Anand Muni was not paying attention or looking at everybody due to his extremely tight schedule. His face was shining as before, and he was blessing everybody with open palm of the right hand. It was indeed genuine blessing from his pure heart. In less than five minutes, he got up and left the hall, accompanied by some followers. Some of the followers were dressed in saintly dress (a white loin cloth and a white sheet called Chadar.) Bijay also followed them in the hope that he would encounter Yogiji (his friend Vinod) somewhere, and request him to help Bijay to get a minute from the busy monk.

Just outside the Hall, and not very far from it, he luckily saw Yogiji greeting the monk with folded hands. Bijay immediately walked towards Yogiji, and bowed to him, who greeted and welcomed Bijay back to the Anand Ashram. He advised him by indication that Bijay must keep quiet for now, as per strict discipline in the Ashram. Bijay kept following Yogiji all the way to Anand Muni's residence. When Anand Muni started having his simple lunch, Yogiji bowed to the monk, and took Bijay along with him to a corner of the holy monk's residence.

Bijay did not want to waste any time. He requested Yogiji to arrange a private one-minute meeting with the holy monk, as he was flying early in the morning next day all the way to USA. Yogiji immediately told him about his helplessness, as that particular day, the holy monk was exceptionally tied up with various important matters regarding Anand Ashram, discussions on various ongoing programs going on in the Ashram, seeing very eminent scholars belonging to different religious groups in the country, abroad etc. etc.

Bijay was concerned that it was not possible for him to solve his curiosity about the monk. Bijay took the courage to ask Yogiji (one time his friend Vinod) as to whether he ever told him about Bijay or mentioned his name to Anand Muni. His answer was strong 'no', as he himself was also curious about the day when the monk called him by his name, Bijay, as if he knew him. Yogiji's intention was to introduce Bijay's entire family, but he never got a chance at that or later time, as the monk was extremely busy, and never paid any attention to Bijay and his family's visit. So Bijay said goodbye to Yogiji, and offered his best wishes to him. He assured him about telling his family in Austin, but Yogiji was indifferent.

Bijay returned to the hotel, but he was restless the whole night. Somehow, he was wrongly or rightly trying to connect the monk to his close friend, Sajjan. That was indeed the first impression and first-time feeling when he saw his picture in Vinod's house in Austin. Now that he saw him face-to-face, he could not correlate Sajjan and the holy monk at all, as there was no similarity of any kind.

Only confusion in Bijay's mind was that how in the world could the monk know his name after all, when he saw him in Anand Ashram.

Bijay had heard a lot of stories about some noble souls capable of knowing a lot more than the ordinary human beings. Anand Muni could very well be such a super soul, and that could be the reason for his calling Bijay by name, even if he did not know him. Bijay could not make himself sure that Anand Muni could ever be his one-time close friend Sajjan, who knew him very well or somebody different.

In the morning, Bijay had breakfast in the hotel with the whole family and started planning for the work that he must do before leaving India for USA. It did not take him much time and he had no problem at all in completing all the formalities, just for example, in paying hotel, food, telephone and laundry bills etc. Then they all decided to go for some shopping in the nearby mall. For the time being, he reconciled within himself not to worry about uncertainties in his mind.

When they all returned to the hotel, it was still early evening, and everybody wanted to take rest. Bijay, however, felt he could go to Anand Ashram quickly, and return by 9 PM or so. Flight was at midnight. He consulted his wife and family about his desire and plan. They all thought that it would be too tight, and if there was unexpected traffic, Bijay would never make it. He had to give up his idea, realizing fully well that he was going too far to solve his problem in the near future. He wished he had planned more time for it, but it was too late!

Bijay left India with his family, as per schedule, and on the plane, he decided to inform his parents about his new dilemma! He knew that his father was too weak to take a fresh step, viz. going to Pune etc. It was out of question for him at his age and in his present physical condition. Even when he was normal, he would never think about going that far for a reason that was not at all convincing! Bijay wished he could locate Sajjan's sister or her address or know about any follower of Anand Muni at Kolkata, but it was only a wishful thinking. However, Bijay did inform his father at Kolkata about Anand Muni at Pune, and his similarity with his close and lost friend Sajjan. Bijay got that gut feeling right on the day he saw his picture in his friend Vinod's house in Austin. He also informed his father about Vinod (his Austin friend) becoming Yogiji Maharaj in Anand Ashram at Pune, completely devoted to his Guru, Anand Muni.

Bijay's father laughed at Bijay's idea of similarity between Sajjan (a family member of Mica-King of India) at Kolkata and Anand Muni (a monk in Pune). He had heard about several monks hypnotizing simple people and misleading them. They are far from spirituality, and have all sorts of addictions, like drinking country liquor, smoking marijuana asking devotees for money etc. He felt that what Bijay's observation from the picture in his friend's house must have been a mere coincidence.

How could a billionaire like Sajjan, married in a very rich family, living in a prestigious complex with his family, turn to monkhood? So when Bijay called him sometime later on, not only his father, but his mother also expressed their opinion that Bijay should not waste his time worrying about worthless people like Anand Muni, as they were convinced that Sajjan could ever be any monk, even if he looked similar to him to Bijay. It was silly to even think about this possibility, according to Bijay's parents at Kolkata.

Time was passing very fast, and Bijay's parents were getting old. They were often sick, and only wished that Bijay and his family stayed with them or at least visited them soon. It so happened that they were attending some relatives's son's wedding at Kolkata, and while they were taking meals, Bijay's father mentioned to his relative (a very spiritual person himself), about Anand Muni at Pune.

His relative immediately remembered a long-time friend, who recently visited him at Kolkata and told him about the miracles and achievements of some relatively unknown Anand Muni. According to him, he was an exceptional monk, and possessed enormous knowledge by his intense meditation, including fasting etc. He was a prolific scholar, and his teachings were remarkable. In fact, he requested his scholar friend at Kolkata to visit him after his son's wedding at Pune. This would enable both of them to go to the beautiful Anand Ashram and retreat, and listen to the holy Anand Muni's unique discourse during his next trip to Pune.

His friend also told Bijay's father that he did go to Pune after his son's wedding, stayed with his friend and all of them with their families went to the beautiful Anand Ashram, which was also a famous picnic spot. All of them had a rare opportunity to listen to a part of Anand Muni's discourse also. That day, he was speaking to the huge audience in the Main Hall on ultimate purpose of human life, that being attaining liberation or salvation. He emphasized that attaining liberation was possible only in human realm, and not in any other realm, like hell, animals, and heaven. Everybody should pay attention to his Karmas or actions, which determine next life. Unless liberation is achieved, a living being passes through as many as 8.2 million cycles of life and death. They were really impressed by his eloquence, yet very simple language. They sincerely wished that they could get some time to see the holy monk face to face, but were told by authorities that he was very busy, and could not see anybody personally without appointment.

Bijay's father was completely surprised that his friend, whom he had seen after a long time, knew about Anand Muni. Both used to meet more often at beautiful Victoria Memorial Gardens at Kolkata for the morning walk. Unfortunately, Bijay's father was now too weak to take a daily walk, but his friend kept close contact with him. In fact, he personally visited him at his residence, and invited him for his son's wedding at Kolkata. As there were a lot of people, who had come for the wedding reception and dinner following it, it was not possible to talk more about Anand Muni at that time.

In about a couple of weeks, Bijay's father called his friend as a courtesy. Besides congratulating him for his son's wedding, he requested him if he knew anything about Anand Muni's pre-monk life at all. It was indeed a very unexpected and undesirable question for his friend, as why should anybody be concerned about the monk's past life? Monks are known from present, as they always concentrate on present only and avoid talking about their past life themselves, nor do they like others to ask or talk about their past life- rather they discourage others. According to them, it was the best way to make progress both for present and future. Still he assured Bijay's father that he himself did not know anything about his pre-monk life. Nevertheless, he would talk to his friend in Pune about it whenever they would talk to each other.

In about a month, Bijay's father got a call from his friend that his Pune friend did not know anything about Anand monk's pre-monk life but assured him that he would try to find out from some of his followers that he knew. He also felt that it was a remote possibility to find out anything about Anand Muni's pre-monk life, unless somebody knew about him before monkhood several years back.

Things were very quiet from all sides. Bijay's father had already informed Bijay about whatever he heard about the holy person. That did not help Bijay. One Sunday morning, Bijay's father's friend invited him to come for dinner to see the Pune friend, who was visiting him. Main reason behind this unexpected invitation was to give Bijay's father an opportunity to talk about the holy monk directly.

When they all were having dinner, the Pune friend himself started talking about the holy man's pre-monk life. His wife knew somebody in the Anand Ashram, who claimed herself to be Anand monk's real sister before his monkhood. She tried to locate her several times in the Ashram celebrations and elsewhere, but did not succeed so far, as she was mostly out of Pune with her family. On asking Anand Muni himself, he had renounced all relationships and he kept completely silent, as if he did not like anybody to talk or ask about his pre-monk life. He never showed any anger, as he was beyond anger, ego, delusion and greed. He emphasized everybody to concentrate on present and present only. She told Bijay's father that she has not given up and would let him know, if she succeeds in locating the apparent so-called sister.

Somehow, out of curiosity, all invitees asked Bijay's father the reason for his finding out about the holy man's pre-monk life. In fact, there are thousands of monks let alone throughout India, in and around Kolkata only. Why was he concerned about Anand monk's pre-monk life alone, and not others? Bijay's father was a bit hesitant to explain his son's madness to locate his lost (out of touch) but closest school friend, Sajjan, who according to Bijay, looked like Anand monk, but Bijay was not sure about his guess. There was a big time-gap of several years since both were in close contact. Bijay's father frankly told them that even he himself has seen Sajjan as an extremely bulky person belonging to a very rich family in Kolkata.

This was the end of that day! Bijay's parents departed after dinner and gave sincerest thanks for their efforts to help Bijay's curiosity (rather madness) and hope for locating his School friend. The Pune friend took the matter very seriously and on return to Pune, he started going to Anand Ashram very often, putting a lot of his time to talk to the holy man's closest followers, Ashram authorities, devoted disciples etc., asking each of them about his pre-monk life. There was no luck so far, and the holy man did not like anybody talking about his past life. He was totally engrossed in his present life only. Time was too precious for him, and he put every minute in helping and guiding others. He did not want anything in return. He was the symbol of complete renunciation and goodwill!

The followers etc. took special care to attend and celebrate the annual foundation day of Anand Ashram. The Ashram was the result of very intensive, devoted and dedicated efforts of many, including the holy monk himself. Besides, being a spiritual place, it was indeed a very beautiful resort-like retreat with all facilities for the tourists, followers, disciples etc. Anybody, irrespective of family status, religious faith, duration and purpose of stay etc. could stay in the Ashram, and everything, including boarding, lodging etc., was free!

It was indeed a heaven for visitors, as it was far from the crowds of Mumbai, Pune etc. It was being supported by charitable donations of several wealthy families, whose members happened to be the key followers of Anand monk. His teachings, way of life etc. touched their hearts, inspired them and transformed their lives. The holy man never asked anybody to become his disciple, nor did he ask anybody for charity. It just came voluntarily by listening to his sermons, looking at his extremely simple life, devoted entirely for the welfare of everybody.

The Pune friend and his wife were on the lookout for the lady, who claimed to be the real sister or at least a close relative of Anand monk. They knew that she and her family lived outside Pune at some place unknown to them, and they were not sure as to whether she had come that day. However, it was such an important occasion that she might be there. The Pune friend, particularly his wife, did not want to miss this rare opportunity of locating this lady, whom she had seen a couple of times in the Ashram on different occasions, but never brought up the issue of the holy man's pre-monk life.

When the ceremony was almost coming to an end, the Pune friend's wife saw that lady (claiming herself to be his sister), standing with folded hands and bowing her head, to Anand monk. She told her husband (Pune friend) immediately and pointed towards her. The objective was that they both must get hold of her that day and that place itself. It was such a big crowd, and there was no way to reach out to her or come closer to her, unless they dared to tear the crowd and pushed people rather rudely and forcefully.

The Pune friend smartly came out of the hall, which had several doors and requested his wife to keep proceeding towards the pedestal near Anand monk, who was blessing everybody with his right hand spreading out to each person. She was rather unethical in crossing and by-passing several people, so that she would not lose the hard-to- get sister-lady. Luckily, the lady was still there, possibly trying to speak to her brother, Anand monk, privately before leaving Anand Ashram.

The Pune friend saw his wife approaching the sought-after-lady. He also decided to go through the huge gathering, pushing the people rather rudely. They both were lucky to get attention from the lady, and politely requested her to spare five minutes for them. The lady agreed and advised them to meet her in the office of Anand Ashram in about one hour.

It was indeed a long wait for them and finally the lady showed up there with her family to see them. The Pune friend thanked her, and politely asked about her relationship with Anand Muni. She emphatically claimed to be his real sister. On being asked about the holy man's pre-monk life several years back, they were told that they belonged to a very well-to-do family from Kolkata. Anand monk renounced the world despite all luxurious and material possessions. He left his wife and two children also in the pursuit of perpetual peace and everlasting bliss. As the lady was getting late, she gave her phone number and address in Mumbai (Bombay) and asked the Pune friend and his wife to call her on phone on next Sunday at ten in the morning.

It was very productive for Pune friend and his wife to have seen and talked to the apparent sister of Anand monk. They surely called her on the following Sunday as scheduled with her. She recognized her conversation with them, and curiously asked about the reason for their enquiry about the pre-monk life of Anand monk. They humbly and frankly told her that it was a request for their close friend at Kolkata, who, in turn, was requested by Bijay's father.

The apparent sister wanted to know more about Bijay's father, but they unfortunately did not know anything about him. At this point, it was not considered proper to talk more about the pre-monk life of the holy man, till they all found about more about Bijay's father and the reason for eagerness to know anything about him. She did emphasize that Anand monk never likes to talk about his past and future. He is always immersed in present, and very conscious of present only. He calls it a stage of 'Bhava Kriya' or spirit of awareness of present and present only. This is the only way to keep mind and body together! He always preaches everybody not to waste the present even for a second. Both mind and body must be concentrated in present and present only.

Message from Pune was duly conveyed to Bijay's father, and eventually to Bijay. Luckily, Bijay now had the phone number and address of the lady claiming to be the sister of Anand monk. He remembered her name, Manisha, and could she be her? He eventually gathered courage to call her long distance from USA to Mumbai (Bombay). Very politely, he introduced himself as a close friend of Vinod (Yogiji) from Austin. He also told her about his futile attempt to find out by himself about Anand monk, when he had his blessings during a trip to Anand Ashram.

On seeing him, he (Anand monk) called him by name i.e. Bijay. How was it possible? Was it just a guess or coincidence or supernatural power that he knew Bijay's name, particularly as Yogiji never mentioned it to him. In fact, Yogiji desperately wanted to introduce his Dallas friend, Bijay, to him, but before that, Anand monk called Bijay by his name, and gave his blessings.

Anand monk was too busy to spare any more time for Bijay. He confirmed his guess that Anand monk's face resembled his one-time closest friend and class-fellow, even though his frail body was far from the rather bulky body of Sajjan. He told her that it was all in Kolkata. After that, Bijay left for USA and lost contact with him for almost ten years. During a trip to Kolkata, his memory was rekindled when he passed through his palatial complex from the airport to his father's residence near Kolkata Club. He made repeated attempts to locate and contact him, even went several times to the palatial mansion where he lived with his family.

On hearing Bijay's story, the lady was convinced that it was a genuine case of a close friend trying to know about Anand monk's pre-monk life. She had never heard about Bijay's name from his brother. She always boasted about knowing everything about his dear brother's pre-monk life, including his accepting the path of monkhood, relinquishing all material possessions, attachments etc. She wanted some time to tell Bijay about it, but before that she felt like consulting Anand monk himself, though she knew he would never talk about it. He had become so pure in thoughts, mind and actions, that he did not ever open his mouth about any question on his personal life- pre or post monkhood. Even if a close follower or devotee asked him about his opinion on any issue or problem, he did not give his opinion to anybody, nor did he ever force anybody to listen to him!

Despite repeated requests, she did not disclose her name to Bijay, as Bijay knew Sajjan's sister's name as Manisha. The lady just put the phone down and disconnected herself, as she had already told Bijay about her checking with Anand monk first regarding disclosing anything about his pre-monk life to him. In fact, she was a bit annoyed with Bijay, asking her repeatedly about Anand Monk being the same as her brother Sajjan pre-monkhood. The question that she had in her mind was what good was it for Bijay to know about it? Where was he when Sajjan, in fact the whole family, particularly Sajjan's wife, Sarita and 2 little children passed through a very, very turbulent life for a long time? She obviously did not wish to ask Bijay at that time, and preferred to get a reaction from the holy monk.

Several months passed by without any call from the lady, claiming to be the real sister of Anand monk. Bijay was a bit disappointed, but having reached this far, he did not wish to give up his efforts to locate his missing close friend Sajjan. During the holidays, he took his family to Austin, and saw Yogiji's wife and family. He requested Vinita, Yogiji's wife, to tell Yogiji, whenever he calls her, about what happened to Bijay's quest for knowing the pre-monk life of the holy man.

Bijay still hoped that as Yogiji is in close contact with Anand monk, he might mention about his Dallas friend, Bijay. How did the holy man know his friend's name, without his ever mentioning about his friend, Bijay? It turned out to be hope only, and Yogiji never got a chance to ask him about this personal matter. The holy man was completely engrossed in spiritual thoughts, and was too busy to listen to anybody about one's personal life! Bijay's father also never heard from his friend at Kolkata about his Pune friend, finding out whatever he had requested long back.

Almost two years were over since the sister lady talked to Bijay, who was now planning a trip to Kolkata, India along with his family to see aging parents on both sides. He made it a point to leave India for USA from Mumbai (Bombay), so that he could visit the beautiful Anand Ashram, and if lucky, with the help of Yogiji, could see holy Anand monk once again. He mustered strength to call the sister lady just to let her know that he planned to see Anand monk during the last week of stay in India. She could not care less, and just ignored his call. In fact, she never talked to Anand monk about Bijay's request to her about knowing the pre-monk life of the holy man!

While at Kolkata, Bijay saw his father's friend and his wife also. He smartly took the address and phone number of their Pune friend. If not anything, Bijay wished to thank them for their kind efforts to help him finding the pre-monk life of the holy man. Bijay and his family were in Kolkata for almost three weeks, visiting relatives on both sides and enjoying shopping etc. Parents on both sides were really very happy to see them once again.

The final week of their departure from Kolkata was approaching fast, and they wished if they could spend some more time with them. As per schedule, Bijay along with his wife and two children departed from Kolkata and reached Mumbai (Bombay) in-route to Dallas, Texas, USA. Nobody in Bijay's own family wished to see the holy monk, and visit Anand Ashram, even though it was beautiful! There were many other worth- seeing places in and around Mumbai, but Bijay was adamant. They made it clear to Bijay that they would rather spend time in shopping malls, having meals in various restaurants, and do some sight-seeing, if possible. Each of them wrongly assumed that Bijay would be with them all the time. Luckily, they had reserved accommodation in a nice hotel near the airport.

First two days, Bijay spent his entire time with rest of the family, and really enjoyed with them. They did not know any relative in Mumbai, and they had all the time for themselves only. So, everybody was happy! On the third day, Bijay thought about the rare opportunity to pursue his efforts about locating Sajjan. He strongly felt in his heart that he must resolve many uncertainties in his mind himself rather than bothering others. He mentioned about his intention to visit the beautiful Anand Ashram again to his wife.

Bijay did not emphasize on Anand monk, but just expressed his desire to see his Austin friend, Vinod, now Yogiji. On getting his wife's consent, he gladly arranged a separate transport for himself, just to take him to Anand Ashram at Pune, stay there for about two hours, and then return back to the hotel. The rest of the family preferred to stay back in the comfortable hotel. After spending most of the day in the hotel, they went to nearby places, mostly walking, and sometimes taking a cab, whenever they needed. The weather was exceptionally nice indeed, and they thoroughly enjoyed.

Bijay reached Anand Ashram, and straight proceeded to the magnificent hall, where Anand monk's sermon was going on, and was about to finish. He could hear him telling the big crowd about the Karmas (deeds). Whatever a living being does, the consequences of one's actions must be borne one way or the other. Ultimate goal of life is to attain liberation, which is possible only by shedding all Karmas done in the past, and to control their ingress in future. First step is to overcome desire, anger, ego and greed.

So intense was the message, coming as light from the holy monk's mouth, that everybody was spell-bound. There was pin-drop silence, but as everyone was seated, Bijay could turn his eyes from one corner of the hall to the other. He could see his Austin friend, now Yogiji, sitting in deep meditation posture, not far from the holy monk. It was impossible for Bijay to walk to him, as he did not want to create any disturbance. He took a seat by himself, anxiously waiting to see Yogiji and possibly the holy monk at the conclusion of the holy serman.

Bijay was just wondering in his mind as to whether he should have first gone to the Pune friend's house and brought them with him. May be that the Pune friend and his wife could help Bijay in finding the lady claiming to be the monk's sister! It was too late to think about all these. Bijay knew it was absolutely the very last chance to do something productive about the long-bothering issue, which made Bijay a laughingstock not only to his own family, but even his parents and their friends and contacts. What will Bijay gain, even if he solves the riddle? It has already cost him a lot of time, money, peace etc., and has created nothing but anxieties, uncertainties etc.

In about half an hour, the sermon was over, and everybody in the big hall stood up to get blessings from the holy monk. Bijay smartly tore the crowd, and proceeded straight to the place, where Yogiji was there, now standing with folded hands like everybody else in the hall. Bijay greeted him politely, and Yogiji, though completely surprised, waived at his Dallas friend, Bijay. Both started following the crowd and when their turn came, the holy man with his radiant face and smile blessed Yogiji and Bijay.

Again, he pointed to Bijay by his name! It was remarkable for him to remember him by name even after so many years. How was it possible? Over the years, he had become weaker, and looked frailer due to intense fasts, extremely busy schedule, concentrated studies, discourses etc. His Anand Ashram was growing every day and was becoming more and more popular to the extent that it became the most desirable picnic spot (retreat) for the families, particularly in the weekend. People, not only from Pune and Bombay, but also from all over India, and in fact all over the world, used to visit the beautiful Ashram, even if they could not see Anand monk personally due to his tight schedule etc.

Bijay got a minute with Yogiji and asked him about the lady, who claims to be the monk's real sister. Luckily, she was there just near the monk and Yogiji pointed out to her. Bijay had never seen her before. He, rather rudely, not even saying goodbye to Yogiji, proceeded fast and straight to the lady, and on reaching there, he very humbly requested the lady for a minute, if possible.

There was a constant influx of people standing quietly in queue with folded hands, hoping to get Anand Muni's Godly Blessings. Though the monk was looking at every person, who passed by him, there was the same feeling of kindness and blessing for everybody. As soon as the last person in the queue passed by, the holy monk, without losing even a second, immediately stood up and started walking very fast towards his residence, where only the close devotees and followers could enter to see him.

Bijay was attentively gazing at the sister lady without any interruption, probably hoping that she would remember his request. At this time, the lady with folded hands, also proceeded towards the monk's residence, probably to say goodbye and also to enquire as to whether everything was going alright in the Ashram, particularly with Anand monk. She knew the very busy schedule of the holy man during the day, and did not want to be an obstacle of any kind to him.

Bijay was becoming impatient. He would obviously be thoroughly disappointed if he missed talking to her. Suddenly, she was kind enough to be looking for Bijay, and asked him to see her just outside one of the entrances to the holy monk's residence. Bijay sincerely followed her instructions, and when she was there to talk to him, Bijay unhesitantly introduced himself as the person from USA, who talked to her on the phone regarding the pre-monk life of the holy man. At that time, somebody in her family, probably her husband, shouted towards her calling her as Manisha again and again.

Bijay distinctly remembered this name as Sajjan's own sister. He did not lose anytime showing gratitude for her time, and requested whether it was possible to see her again, wherever and whenever convenient, before his flight to USA on early morning of coming Thursday. Bijay left Kolkata on last Thursday night, and spent Friday and Saturday with his family. He knew from his previous visit that special events are usually performed in the Ashram on Sundays only, as majority of Anand monk's followers work on other days, even though the Ashram is always open. The holy man conducts special and very intense spiritual courses almost every day other than Sundays, and most retired devotees attend these.

While Manisha was walking towards the car, where her husband was waiting, Bijay did not lose anytime introducing himself as a close school friend of Sajjan at Kolkata, and he himself telling Bijay about his only younger sister, Manisha, who was studying at La Martier School at that time.

Manisha had apparently talked to Anand monk about somebody by the name of Bijay enquiring about his pre-monk life. Anand monk did not care one way or the other and did not show any emotion or enthusiasm. As he did not say anything otherwise, Manisha took it as a signal to allow her to do whatever pleases her. So she asked Bijay to see her at her Merryman Point residence on Wednesday at 9 PM after dinner only for 10 minutes. She gave Bijay her residence address. Thus, Bijay felt that it was the most productive day for him.

Bijay followed Manisha's instructions. He reached Manisha's beautiful residence in the posh area of Bombay, i.e. Merryman Point, at 9 PM sharp, and profoundly thanked her very humbly and sincerely for all the time she was giving to Bijay. Manisha herself started telling him that they belonged to a very well-to-do, aristocratic family, and at one time, were called 'Mica Kings Of India'. She herself and her brother, now Anand monk, used to live in one of the palatial houses in a big Complex in one of the most coveted areas of Kolkata along with their parents. They lost their mother first, and then the father, when they both were still young.

Business was being run on father's death by Sajjan, who was still learning and was not prepared to take this big responsibility. His young father's sudden death was a big calamity for him. Sajjan's cousin brothers had their own businesses and they used to live in their own houses in the spacious Compound. There was nothing to do between Sajjan and them so far as individual businesses, taxes etc. were concerned. They rarely saw each other. Everything was apparently going smooth for several years since Sajjan took over his part.

Suddenly, Sajjan got a registered notice from the Federal Income Tax Division of the Government Of India. It said that he has not paid his taxes, which had accumulated to a big amount with interest, penalty etc. He was summoned to appear at the Tax Office with the only option of paying the entire amount in 15 days from the date of the letter.

Sajjan was truly shocked, as he used to pay taxes for his business enterprise every year in time. As the tax returns were prepared and filed by his CPA, he contacted his CPA immediately. The CPA could not understand the Tax Notice at all, and he wanted to see the Notice himself. He rushed straight to Sajjan's house, where Sajjan was desperately waiting for him. In spite of his expertise as a reputed CPA, he was completely lost to see the huge arrears to be paid by Sajjan as per Notice. He assured Sajjan that he would find out the details next morning. He took Sajjan along with him and his staff. On reaching the Tax Office, as the CPA knew the tax authorities, he walked straight to the relevant office. He showed the year-by-year tax payments by Sajjan as per the Federal Tax Returns prepared by him. He point blank asked the validity of this Notice to him.

Was it sent by mistake? Was it intended for somebody else? The CPA challenged them and wanted an immediate answer, so that his client Sajjan could be relieved of this unexpected anxiety. He also assured them that, if justified as per Tax Rules, Sajjan certainly would pay any unpaid tax related to his business enterprise immediately.

The Tax Officers opened all files and found out that Sajjan's cousin brothers, in spite of having their own and separate enterprises, took advantage of Sajjan's simplicity, innocence and immaturity. They all jointly and quietly transferred their tax obligations to him without his knowledge. They never got his consent, and never ever gave even a penny to him, nor did Sajjan ever expect anything from their businesses. There was no document showing Sajjan having any interest in any of their businesses. Their tax returns were also there, but tax due in each case was linked to Sajjan by fraudulent means.

At that stage, Sajjan and his CPA saw each cousin one by one and got the blunt reply that Sajjan's father had legally committed in writing to pay the entire taxes for them. They all promised to show Sajjan the document apparently signed by his father long time ago. Sajjan's CPA asked each of them that in that situation, Sajjan's father must have been a partner in each enterprise, and they must see the agreement made by Sajjan's father, before proceeding further. Sajjan's father never ever mentioned to him, nor did Sajjan get any partnership return from cousins etc. How could the entire taxes be his father's responsibility? There was no legal way to understand the huge tax burden on Sajjan's father, unless he had substantial partnership in each enterprise owned by Sajjan's cousin brothers. There was no partnership agreement and there was no tax return filed for these partnerships.

It was getting <u>late</u> for Manisha as the whole family was waiting for her to finish talking to Bijay, whom they did not know. Her husband was already furious and was blowing the horn in the car. He had to prepare for the next day and the children had to go to school. Even Manisha had to do laundry etc.

Bijay also had to return to the hotel, as he had early morning flight. Needless to say, he was still at a loss as to figure out Sajjan's transformation from the life of a business tycoon to a holy man's life. He was also shocked and surprised to know that his innocent friend was probably dragged into fraud and manipulations by his own family members. He got the feeling that Sajjan was such a simple and straight-forward person that he did not or could not do enough to reverse the complex situation and get his rightful share. Whatever it was, they all had to leave the conversation at that stage only. Bijay thanked Manisha, who voluntarily asked Bijay to call her from USA, if he had any more question.

Bijay flew to Dallas, and he was now a lot more relaxed. He had to take care of a lot of pending things, including paying bills, his and his wife's work, children's studies, including homework etc. before worrying about calling Manisha and finding out more from the point she left. Once things appeared to be settling down normally, one morning on a weekend, he called Manisha and wondered as to whether she was available to answer some more questions about Sajjan.

Manisha told Bijay that as nothing came out from personal meetings with his cousins etc., Sajjan and his CPA decided to take the matter to the court and vehemently rejected the idea of paying taxes on businesses not at all related to Sajjan's business interests. As per CPA's suggestion, an expert attorney was hired to fight the case for Sajjan. It really cost Sajjan a lot. Weeks after weeks, the matter did not move further in the court, but taxes and attorney's fees got compounded day by day.

A lot of irrelevant expenses were linked to Sajjan and his father by the close relatives by manipulations and fake receipts. It was indeed a big trap to come out of it. Sajjan was disappointed, frustrated and lost a lot of weight. He was very, very sick for a number of days, and started losing interest in his responsibilities towards his business, his wife, children and close relatives, who were still on his side. His attorney and CPA tried their best, but nothing was moving in Sajjan's favor. All documents were false, fake and manipulated- it was indeed a monstrous fraud for many years. Sajjan could not see even a ray of hope.

On the other side, he started taking interest and putting effort in Yoga, meditation, spiritual books, spiritual discussions and meetings etc. This resulted in a big mental, though yet silent, transformation in Sajjan's attitude. This transformation was too rapid to be reversed. And one day, he decided to renunciate everything, including his family. Nobody could trace him out anywhere for months after months, despite all efforts. As he could not be traced anywhere, his family, consisting of his wife and two sons, decided to move to his wife's parents living in Meerut near Delhi.

All of them, including Manisha, lost hope of finding him till one day, they saw a big ad in the Kolkata newspaper Statesman about seizing the Palatial Complex, and putting on auction. They all figured out that it must be due to huge taxes which Sajjan could not, and did not pay, despite several reminders from the Income Tax Authorities over the years. Sajjan's cousins and their families were forced to evacuate the Complex, and they all moved to different places, probably rejoicing over their misdeeds, fraudulent and malicious actions towards the innocent and simple Sajjan and his family members. They still did not have any idea on whereabouts of Sajjan and his new life. Nobody imagined at that time that his penance would be so immense, deep and exhaustive reaching a point of no return to family life!

Manisha, one day, while glancing over the newspapers, came across an attractive picture in a local paper of Bombay (Mumbai), showing a very peaceful, spiritual, and yet exceptionally beautiful and scenic resort in Pune. Her family, including herself, decided to visit the retreat one long weekend. They had heard from several friends about the sermons of a holy monk over there. In fact, some of these friends were his devotees and followers, who travelled all the way from

Mumbai to Pune just to listen to him. When Manisha and family went there, they all were walking in the beautiful resort to find a suitable restaurant, where they could have some snacks, tea etc. As they were walking, they happened to hear the special sermon of the noble monk coming from a very big assembly Hall, not very far from there.

Nobody in Manisha's family, including herself, was interested to hear the monk, as they all came to the retreat just for fun and relaxation. So they just decided to return to the car parked on the outskirts of the retreat. The resort was called Anand Ashram, and was really very clean, scenic, well-organized and well-managed. A passerby handed over a leaflet to Manisha just as a token of giving information to the visitors about the holy monk's special sermon that day. When Manisha looked at the leaflet while walking towards the car along with her family members, it had Anand monk's picture as an inset. She was startled to see the holy man's face, which very much resembled her missing brother Sajjan's face. She requested her family members, particularly her husband, to wait for her by the hall on way to the car parking place, and though unwilling, he just agreed seeing no other option.

They saw a huge crowd of people with folded hands bowing to the monk. Manisha could not resist her curiosity to know more about the holy man, particularly as his face resembled that of his own elder brother, Sajjan. Only difference was the unmatched and unthinkable glow (aura) around his face, as shown in the leaflet. Also, Sajjan was very fat, but in the leaflet, he looked slim. She wanted to see the monk with her own eyes, as from the picture, he appeared to be a divine entity with all happiness, respect, bliss and love around him.

Manisha realized that her family was desperate in her returning back to the car. So she forced herself through the crowd and moved close to the monk with folded hands. When Anand monk saw her, he, without showing any emotion or excitement, called her by her name - Manisha. Manisha was overwhelmed with unprecedented joy, as if she was dreaming! How could he call her by name, unless he was and is Sajjan, her own brother? He showed absolutely no emotion or excitement on seeing her.

Manisha rushed to the car and informed her family desperately waiting for her to return, about the strange feelings towards the holy monk, popularly known as Anand monk. Her family just could not believe her, as her brother, Sajjan, could not be traced anywhere despite so many efforts. The Income Tax authorities were after him, and they had informed Media, Police etc. Having given up any hope of finding him, they finally had a Public Auction of the huge Palatial Complex. And today, just by stroke of luck, she may have found her brother! It was too hard to be true for her and her family!

First thing Manisha did on reaching home was to inform his brother's wife, Sarita, who had moved to her parents' home in Meerut. Just like Manisha, she could not believe it! She was overwhelmed and wanted to see him personally. In her mind, she was confident that on seeing him and spending some time with him, she would bring him back to a normal householder's life, giving up monkhood. Her family suffered so much due to Sajjan's sudden disappearance due to fraud, deception and manipulations by his cousins. Everybody, including her family at Meerut, gave up on ever finding him. She requested Manisha to keep it a secret, as the news of finding him would be disastrous for miscreants, though they all had powerful attorneys, fake documents etc. to protect them in any circumstance.

Manisha totally agreed with her, and immediately cautioned her husband and children not to talk to anybody. She wanted to visit the Anand Ashram again and have a glimpse of the holy monk a little more closely. Only on confirming that Anand Muni was her own brother Sajjan, she would invite others, only the very close, including her sister-in- law Sarita (Sajjan's wife). They both agreed, though they were extremely anxious to unfold the mystery- too good to be true!

Manisha was requesting her husband all the time to visit Anand Ashram again. She was eager to know the date of a particular Sunday when her family could go and have a nice picnic in the beautiful resort. She internally wished in her heart that while rest of the family members took food followed by rest in the cool and gentle breeze, she would rush straight to the holy monk tearing the huge crowd anxious to have his own brother's glimpse with blessings immediately after his special sermon.

Manisha wished to hear Anand monk calling her by name once again. This would pacify all her doubts about his being her own dear elder brother Sajjan. Her husband finally agreed in a fortnight to go there on a Sunday. Manisha and her family were really glad for his consent to drive down to Anand Ashram. Manisha called the Ashram authorities to confirm that there was a special sermon that day, including expected time for the holy monk to conclude his sermon. Everything appeared to be going well.

Quite unexpectedly, one of the authorities called Manisha just when they were ready to leave their house in Mumbai for the Ashram in Pune, informing her that the holy man's program was cancelled by himself due to unknown reasons. This was a very unpleasant and completely unexpected message for Manisha. Immediately, her mind went to Sarita Bhabhi (Sajjan's wife) and others in Meerut. What would she tell them? She herself wanted to know the reason for cancelling the holy man's program.

Was the completely unexpected and sudden cancellation due to the holy monk, not feeling well? In that case, she wished she could be with him, her own elder brother Sajjan in real life. Was it due to his extreme exhaustion? When she saw him last, he appeared well despite his round-the-clock spiritual discourses. How is she going to get answers to her questions? It came to her mind that she should request one of the authorities of the Ashram to call her once something was known about him.

Manisha promptly called the Ashram Office and was assured of immediate response in case of some information about the holy man. She promptly informed Sarita Bhabhi (Sajjan's wife) about these unexpected developments, and requested her not to worry and wait for her call. There was no point in coming to Pune, as she would not be able to see the holy monk (her husband) without his consent.

Days after days, weeks after weeks passed, but there was no message at all from the Ashram. Manisha kept on reminding them every time. At heart, she was extremely worried, and regretted that she could not be near her own brother, the holy Anand monk. There were repeated phone calls from Sarita, who insisted on coming to Pune. She, somehow, felt confident she would be able to break this long silence. Manisha, though she would surely welcome Sarita, she knew that Ashram authorities would never allow her to see or go near her husband without his consent. They themselves were desperate to know about the holy man, the reason for cancelling all programs, his health etc.

Only Yogiji, the monk's closest and most trusted devotee and disciple, was allowed to go inside just to keep food on the table, but he was under oath from the holy man to come out immediately without any disturbance. He could neither see him, nor talk to him. He could not find out anything about his Guru, and was getting more and more concerned every day. He was closest to him, but he did not wish to disobey his instructions about no disturbance at all!

Matter was getting very, very serious for Ashram authorities, the holy monk's hundreds of disciples, Yogiji, Manisha, Sarita and several others. They all felt that they deserved to know more about the holy monk, his plans etc. Question in everybody's mind was as to when programs, particularly his sermons, would be resumed again? They all waited too long, and wished to serve the Ashram and the holy monk with whatever was needed! They all decided that Yogiji Maharaj,when he would go next time to serve food to Anand monk, should speak to the holy monk about his health, plans etc., and offer him anything that he wanted.

Yogiji, as usual, went to deliver the food, but he did not come out immediately. He had the obligation of finding out more about the holy monk, who was till now keeping his disciples and followers in absolute suspense about his plans etc. He had to do so in order to perform intense meditation. That day, he was rather disappointed to see his closest disciple violate his instructions.

Still the monk blessed him, when he saw him with folded hands, and immediately went in deep meditation without taking food. He wanted to give up any attachment to his body. His sermons used to be about the Science of difference between body and soul. Soul is immortal, but body always perishes. His food was lying there, but it looked like he was taking only water.

Yogiji was still standing there with folded hands in the hope of hearing from him, but the holy monk went back in deep meditation. He was so concentrated that it was obvious that he had no attachment of any kind with outside world, including his own Anand Ashram, his frail body, getting weaker and weaker. Several hours went by, and Yogiji desperately gave up following on him. Next day, when he went with food, he saw the entire food left by him a day earlier. Only a little bit of water was used. He could not see the holy monk anywhere.

Days after days went by with no luck. There was complete silence from the holy monk. In about 3 weeks, he stopped drinking water also, and was observing complete fast. Yogiji informed everybody. They all knew that the end was near. Nobody was allowed to disturb him, as he was too weak even to stand or sit. He was just lying down on the floor, covering himself with the white chadar. He did not ask for any help. His sister Manisha was informed by the Ashram authorities. Manisha informed Sarita. They were all helpless and realized the holy monk was at a point of no return. He was no longer a body but was an immortal soul. He was absolutely detached and was on his way to a journey towards eternity.

When he was gone forever, all his devotees gathered to pay their respects to a noble person, who sacrificed his life for them. Why did he do that? Why did not he continue to give sermons? Yogiji, his nearest and closest disciple, took control of the situation. He politely told them that he took everybody's permission to proceed on to deep meditation. He also emphasized on his sermon for utilizing this rare opportunity of belonging to human realm in attaining liberation. In order to achieve later, one must eradiate all attachments and aversions. Nothing remains at that stage except pure soul!

9 798889 250586